MW00800106

Industrial Methods for the
EFFECTIVE DEVELOPMENT AND TESTING
OF DEFENSE SYSTEMS

Panel on Industrial Methods for the Effective Test
and Development of Defense Systems

Committee on National Statistics
Division of Behavioral and Social Sciences and Education

Board on Army Science and Technology
Division on Engineering and Physical Sciences

NATIONAL RESEARCH COUNCIL
OF THE NATIONAL ACADEMIES

THE NATIONAL ACADEMIES PRESS
Washington, D.C.
www.nap.edu

THE NATIONAL ACADEMIES PRESS   500 Fifth Street, N.W.   Washington, DC 20001

NOTICE: The project that is the subject of this report was approved by the Governing Board of the National Research Council, whose members are drawn from the councils of the National Academy of Sciences, the National Academy of Engineering, and the Institute of Medicine. The members of the committee responsible for the report were chosen for their special competences and with regard for appropriate balance.

Support for the work of the Committee on National Statistics is provided by a consortium of federal agencies through a grant from the National Science Foundation (grant number SES-0453930). The project that is the subject of this report was supported by an allocation of the National Science Foundation by the U.S. Department of Defense under this grant. Any opinions, findings, conclusions, or recommendations expressed in this publication are those of the author(s) and do not necessarily reflect the view of the organizations or agencies that provided support for this project.

International Standard Book Number-13:   978-0-309-22270-9
International Standard Book Number-10:   0-309-22270-2

Additional copies of this report are available from the National Academies Press, 500 Fifth Street, N.W., Lockbox 285, Washington, DC 20055; (800) 624-6242 or (202) 334-3313 (in the Washington metropolitan area); Internet, http://www.nap.edu.

Suggested citation: National Research Council. (2012). *Industrial Methods for the Effective Development and Testing of Defense Systems.* Panel on Industrial Methods for the Effective Test and Development of Defense Systems. Committee on National Statistics, Division of Behavioral and Social Sciences and Education and Board on Army Science and Technology, Division on Engineering and Physical Sciences. Washington, DC: The National Academies Press.

THE NATIONAL ACADEMIES
Advisers to the Nation on Science, Engineering, and Medicine

The **National Academy of Sciences** is a private, nonprofit, self-perpetuating society of distinguished scholars engaged in scientific and engineering research, dedicated to the furtherance of science and technology and to their use for the general welfare. Upon the authority of the charter granted to it by the Congress in 1863, the Academy has a mandate that requires it to advise the federal government on scientific and technical matters. Dr. Ralph J. Cicerone is president of the National Academy of Sciences.

The **National Academy of Engineering** was established in 1964, under the charter of the National Academy of Sciences, as a parallel organization of outstanding engineers. It is autonomous in its administration and in the selection of its members, sharing with the National Academy of Sciences the responsibility for advising the federal government. The National Academy of Engineering also sponsors engineering programs aimed at meeting national needs, encourages education and research, and recognizes the superior achievements of engineers. Dr. Charles M. Vest is president of the National Academy of Engineering.

The **Institute of Medicine** was established in 1970 by the National Academy of Sciences to secure the services of eminent members of appropriate professions in the examination of policy matters pertaining to the health of the public. The Institute acts under the responsibility given to the National Academy of Sciences by its congressional charter to be an adviser to the federal government and, upon its own initiative, to identify issues of medical care, research, and education. Dr. Harvey V. Fineberg is president of the Institute of Medicine.

The **National Research Council** was organized by the National Academy of Sciences in 1916 to associate the broad community of science and technology with the Academy's purposes of furthering knowledge and advising the federal government. Functioning in accordance with general policies determined by the Academy, the Council has become the principal operating agency of both the National Academy of Sciences and the National Academy of Engineering in providing services to the government, the public, and the scientific and engineering communities. The Council is administered jointly by both Academies and the Institute of Medicine. Dr. Ralph J. Cicerone and Dr. Charles M. Vest are chair and vice chair, respectively, of the National Research Council.

www.national-academies.org

PANEL ON INDUSTRIAL METHODS FOR THE EFFECTIVE TEST AND DEVELOPMENT OF DEFENSE SYSTEMS

VIJAY NAIR (*Chair*), Department of Statistics and Department of Industrial and Operations Engineering, University of Michigan

CHARLES E. (PETE) ADOLPH, Independent Consultant, Albuquerque, NM

W. PETER CHERRY, Science Applications International Corporation, Ann Arbor, MI (Retired)

JOHN D. CHRISTIE, Logistics Management Institute, Alexandria, VA

THOMAS P. CHRISTIE, Independent consultant, Arlington, VA

A. BLANTON GODFREY, College of Textiles, North Carolina State University

RAJ KAWLRA, Manufacturing Quality, Chrysler LLC, Auburn Hills, MI

JOHN E. ROLPH, Department of Industrial Operations and Management, Marshall School of Business, University of Southern California

ELAINE WEYUKER, AT&T Laboratories, Florham Park, NJ

MARION L. WILLIAMS, Institute for Defense Analyses, Alexandria, VA

ALYSON G. WILSON, Science and Technology Policy Institute, Institute for Defense Analyses, Washington, DC

MICHAEL L. COHEN, *Study Director*

MICHAEL J. SIRI, *Program Associate*

COMMITTEE ON NATIONAL STATISTICS
2011-2012

Acknowledgments

The Panel on Industrial Methods for the Effective Test and Development of Defense Systems expresses its appreciation to the many individuals who provided valuable assistance in producing this report. We appreciate the support of Michael Gilmore, Director of Operational Test and Evaluation (DOT&E), and Frank Kendall, Principal Deputy Under Secretary of Defense (Acquisition, Technology, and Logistics) at the U.S. Department of Defense (DOD). We are also greatly indebted to Nancy Spruill, director, Acquisition Resources and Analysis, and Ernest Seglie, recently retired science advisor to the director of operational test and evaluation at DOD.

The success of the study depended greatly on the presentations at the workshop, which was the panel's main fact-finding activity. The panel is extremely grateful to the major speakers who represented industry perspectives: Donald Bollinger, Hewlett-Packard; Salim Momin, SRS Enterprises; Sham Vaidya, IBM; and Jeffrey Zyburt, DCYI Engineering Consulting & Development Process. The other presentations by Michael Cushing, Army Evaluation Center, and Robin Pope, SAIC (Science Applications International Corporation), were also very helpful in preparing this report. We also thank Steve Hutchinson, Defense Information System Agency, DOD, and Dmitry Tananko, General Dynamics, who served as discussants. The workshop also included a very productive panel session with reactions of the defense test community to the presentations by the experts from industry, and we thank them: William McCarthy, DOT&E; Steve Welby, director, Systems Engineering; and Chris DiPetto, acting director, Development Test.

ACKNOWLEDGMENTS

We also thank the staff of the Committee on National Statistics, especially Michael Siri and Anthony Mann, for their smooth organization of our meetings, and Julie Schuck for her work on the report draft and for helping plan and support the panel's meetings. We would also like to express our gratitude to Eugenia Grohman for the technical editing of the report.

Most importantly, we were fortunate to have an outstanding group of colleagues on the panel. They provided critical insights and expertise on industrial processes and systems engineering as well as defense acquisition and testing. They volunteered their time and service generously before, during, and after the panel meetings and were involved extensively in the writing of the report.

This report has been reviewed in draft form by individuals chosen for their diverse perspectives and technical expertise, in accordance with procedures approved by the Report Review Committee of the National Research Council (NRC). The purpose of this independent review is to provide candid and critical comments that will assist the institution in making its published report as sound as possible and to ensure that the report meets institutional standards for objectivity, evidence, and responsiveness to the study charge. The review comments and draft manuscript remain confidential to protect the integrity of the deliberative process. We thank the following individuals for their review of this report: Donald Bollinger, distinguished technologist, Hewlett-Packard; Gilbert F. Decker, consultant, Los Gatos, CA; Arthur Fries, staff member and project leader, Institute for Defense Analyses; Charles E. McQueary, consultant, former under secretary for science and technology, U.S. Department of Homeland Security and former director of operational test and evaluation; Department of Defense, Greensboro, NC; William Meeker, Department of Statistics, Iowa State University; Duane Steffey, director, Statistical and Data Sciences, Exponent®, Menlo Park, CA; Dmitry Tananko, manager, Reliability, General Dynamics Land Systems; and Jeff Zyburt, president, DCYI Consulting, New Hudson, MI.

Although the reviewers listed above have provided many constructive comments and suggestions, they were not asked to endorse the conclusions or recommendations nor did they see the final draft of the report before its release. The review of this report was overseen by Thom J. Hodgson, distinguished university professor, Fitts Industrial and Systems Engineering Department, North Carolina State University Appointed by the NRC's Report Review Committee, he was responsible for making certain that an independent examination of this report was carried out in accordance with institutional procedures and that all review comments were carefully considered. Responsibility for the final content of this report rests entirely with the authoring panel and the institution.

The panel recognizes the many federal agencies that support the Committee on National Statistics directly and through a grant from the National Science Foundation. Without their support and their commitment to improving the national statistical system, the work that is the basis of this report would not have been possible.

Vijay Nair, *Chair*
Michael L. Cohen, *Study Director*
Panel on Industrial Methods for the Effective
Test and Development of Defense Systems

Contents

Glossary and Acronyms

ACAT: Acquisition category, a designation for each defense program based on program costs that determines both the level of review that is required by law and the level at which *Milestone decision authority* rests in DOD.

ACAT I: Of four acquisition categories (ACAT I to ACAT IV), the most expensive systems, which are estimated to require either more than $365 million (fiscal 2000) for research and development or more than $2.19 billion (fiscal 2000) for purchase of the specified number of delivered systems.

Defense Acquisition Board (DAB): A senior advisory board for defense acquisitions in DOD that includes the vice chairman of the Joint Chiefs of Staff and the service secretaries, among others, and that plays a key role since it is responsible for approving major defense acquisition programs.

Developmental test (and evaluation): Typical testing of a defense system early in development, analogous to laboratory or bench testing, sometimes involving only components or subsystems, that often does not represent full operational realism, in contrast with *Operational test (and evaluation).*

Director, Cost Assessment and Program Evaluation (CAPE): The principal staff assistant to the secretary of defense for cost assessment and program evaluation, whose responsibilities include analysis and evaluation of plans, programs, and budgets in relation to U.S. defense objectives, projected threats, allied contributions, estimated costs, and resource constraints and ensuring that the costs of DOD programs, including classified programs, are presented accurately and completely.

Director, Defense Research and Engineering (DDR&E): The principal staff adviser to *USD-AT&L* for matters of research and engineering.

Director, Operational Test and Evaluation: The office or the person who heads *DOT&E.*

DOT&E: Office of the Director of Operational Test and Evaluation (or, sometimes, the person who holds the office), a unit in the Office of the Secretary of Defense, which also reports directly to Congress, responsible for DOD policies and procedures for analyzing and interpreting the results of operational testing and evaluation for each major DOD acquisition program, approving test plans, and providing independent evaluations of *ACAT I* systems.

Effectiveness and suitability[1]: A measure of the overall ability of a system to accomplish a mission when used by representative personnel in the environment planned or expected for operational employment of the system considering organization, doctrine, tactics, supportability, vulnerability, and threat. Effectiveness is the degree to which a system can carry out its mission when fully operational. (Operational) suitability is the degree to which a system can be placed and sustained satisfactorily in field use.

Evolutionary acquisition: The development of a defense system in stages, with the systems that result from each stage of development potentially released to the field.

5000.01: DOD directive that provides management principles and mandatory policies and procedures for managing all acquisition programs.

5000.02: DOD instruction that establishes a simplified and flexible management framework for translating capability needs and technology opportunities.

---

[1]Definition adapted from Joint Capabilities Integration and Development System, CJCSI 3170.01G. See http://jitc.fhu.disa.mil/jitc_dri/pdfs/3170_01g.pdf [December 2011].

Full-rate production: The final step of procurement, in contrast to release to the field of a small number of units as part of low-rate initial production, which requires either the judgment that it is effective and suitable by *DOT&E* or by a full-rate production decision review.

HP-UX: Hewlett-Packard's implementation of the UNIX operating system.

Initial Operational Test and Evaluation (IOT&E): The first large operational test of a system or system element [see *Operational test (and evaluation)*].

Joint Capabilities Integrated Development System (JCIDS): A formal DOD procedure that defines requirements and evaluation criteria for defense systems in development.

Materiel developer: The organization or command responsible for providing materiel to DOD or specific service forces, with responsibilities that include research and development of weapon systems.

Milestone A: The step in the *Milestone system* that promotes a system to the technology development phase of development.

Milestone B: The step in the *Milestone system* that promotes a system to the engineering and manufacturing development phase of development.

Milestone decision authority: The person or office responsible for the decision to promote a system to the next step of development in the *Milestone system*.

Milestone system: A set of three milestones that bridge the four steps of defense acquisition: (1) materiel solution analysis, (2) technology development, (3) engineering and manufacturing development, and (4) production and deployment.

Model-based engineering: Systems engineering, starting from development of requirements, through development of components and subsystems, then integration into full systems, that is guided throughout by the use of models that simulate overall system performance of systems comprised of various kinds of subsystems and components, which enforces collaboration across multiple engineering departments.

Modeling and simulation: Various methods for simulating, sometimes with system components in the loop and sometimes entirely computer based, the functioning of a (proposed) defense system.

Operational test (and evaluation): Testing of a defense system relatively late in development, involving the full system in whatever numbers will be used cooperatively in the field, in scenarios that attempt to represent full operational realism, including representation of enemy systems, countermeasures, and operated by users with training typical of fielded systems.

Program Management Office (PMO): The office tasked with development, production, and sustainment of a defense system on a timely basis that satisfies a set of requirements at a given price.

Program manager[2]: The person with responsibility for and authority to accomplish program objectives for development, production, and sustainment to meet the user's operational needs and accountable for credible cost, schedule, and performance reporting to the *Milestone decision authority*.

Reliability, Availability, and Maintainability (RAM)[3]: The probability of an item to perform a required function under stated conditions for a specified period of time (reliability), degree to which it is in an operable state and can be committed at the start of a mission when the mission is called for at an unknown (random) point in time (availability), and its ability to be retained in, or restored to, a specified condition when maintenance is performed by personnel having specified skill levels, using prescribed procedures and resources, at each prescribed level of maintenance and repair (maintainability).

Technology readiness level: The degree to which the behavior of a newly developed technology is understood well enough for incorporation into a system in *Full-rate production*.

---

[2]Definition adapted from the U.S. Department of Defense Directive 5000.01. See http://www.dtic.mil/whs/directives/corres/pdf/500001p.pdf [December 2011].

[3]Definition adapted from the U.S. Department of Defense Guide for Achieving Reliability, Availability, and Maintainability. See http://www.acq.osd.mil/dte/docs/RAM_Guide_080305.pdf [December 2011].

Test and Evaluation Master Plan (TEMP): A formal document that provides a scheme to be used to create detailed test and evaluation plans, especially schedule and resource commitments.

U.S. Army Training and Doctrine Command (TRADOC): An Army element that provides training to soldiers and, as part of that training, helps design, develop, and integrate new capabilities and doctrine.

USD-AT&L: The Under Secretary of Defense for Acquisition, Technology, and Logistics, the primary office in the Office of the Secretary of Defense responsible for the development and acquisition of defense systems.

Summary

This report responds to a request from the U.S. Department of Defense (DOD) to identify engineering practices that have proved successful for system development and testing in industrial environments. It is the latest in a series of studies by the National Research Council (NRC), through the Committee on National Statistics, on the acquisition, testing, and evaluation of defense systems. The previous studies have been concerned with the role of statistical methods in testing and evaluation, reliability practices, software methods, combining information, and evolutionary acquisition. This study was sponsored by DOD's Director of Operational Test and Evaluation (DOT&E) and the Under Secretary of Defense for Acquisition, Technology, and Logistics (USD-AT&L). It was conducted by the Panel on Industrial Methods for the Effective Test and Development of Defense Systems.

The study panel's charge was to plan and conduct a workshop to explore how developmental and operational testing, modeling and simulation, and related techniques can improve the development and performance of defense systems, particularly techniques that have been shown to be effective in industrial applications and are likely to be useful in defense system development. This workshop was the panel's main fact-finding activity, which featured speakers who described practices from software and hardware industries.

We emphasize that we could not, and did not, carry out a comprehensive literature review or examination of industrial and engineering methods for system development. Rather, drawing on information from the workshop and the experience and expertise of the panel's members, we focused on the techniques that have been found to be useful in industrial system development and their applicability to the DOD environment, while acknowledging the differences in the two environments. To that end, we also considered the availability and access to data (especially test data), the availability of engineering and modeling expertise, and the organizational structure of defense acquisition.

Many, perhaps even most, of the industrial practices we discuss and recommend are or have been used in DOD, but they are not systematically followed. We do not offer new policy or procedural recommendations when (1) the techniques are already represented in DOD acquisition policies and procedures, (2) DOD has been trying to implement the desirable practices, or (3) the desirable practices have previously been recommended in other NRC reports or by other advisory bodies. In these cases we reiterate the benefits of and the need to fully adopt and follow the relevant policies, procedures, and practices. We do offer recommendations to determine if the defense acquisition community is moving in the wrong direction by reviewing policies, procedures, and practices that are new or have elements that are new.

REQUIREMENTS SETTING

Conclusion 1: It is critical that there is early and clear communications and collaboration with users about requirements. In particular, it is extremely beneficial to get users, developers, and testers to collaborate on initial estimates of feasibility and for users to then categorize their requirements into a list of "must haves" and a "wish list" with some prioritization that can be used to trade off at later stages of system development if necessary.

Although communication with users is common in defense acquisition, the emphasis at the workshop was on a continuous exchange with and involvement of users in the development of requirements. In addition, the industrial practice of asking customers to separate their needs into a list of "must haves" and a "wish list" forces customers to carefully examine a system's needs and capabilities and any discrepancies between them and thus make decisions early in the development process. It is also important to use input from the test and evaluation community in the setting of initial requirements.

Conclusion 2: Changes to requirements that necessitate a substantial revision of a system's architecture should be avoided as they

can result in considerable cost increases, delays in development, and even the introduction of other defects.

Having stable requirements during development allows the system architecture to be optimized for a specific set of specifications, rather than be modified in a suboptimal manner to try and accommodate various updates and changes over time. However, there must also be some flexibility that allows for modifications that are responsive to users' needs and changing environments. Although existing DOD regulations mandate that changes in requirements must go through a rigorous engineering assessment before they are approved, these regulations do not appear to be strictly enforced.

Conclusion 3: Model-based design tools are very useful in providing a systematic and rigorous approach to requirements setting. There are also benefits from applying them during the test generation stage. These tools are increasingly gaining attention in industry, including among defense contractors. Providing a common representation of the system under development will also enhance interactions with defense contractors.

The term "model-based design tools" relates to formal methods used to translate and quantify requirements from high-level system and subsystem specifications, assess the feasibility of proposed requirements, and help examine the implications of trading off various performance capabilities (including various aspects of effectiveness and suitability, including durability and maintainability). It has also been called model-based engineering. In addition to rigorously assessing the feasibility of proposed requirements and helping assess the results of "lowering" some requirements while "raising" others, model-based design tools are known to provide a range of benefits: a formal specification of the actual intent of the functionality, they document the requirements; the model is executable, so any ambiguities can be identified; the model can be used to automatically generate test suites; and, possibly most importantly, the model captures knowledge that can be preserved.

DOD should have expertise in these tools and technologies and use them with contractors and users. More broadly, DOD should actively participate, if not lead, in the development of model-based design tools.

Recommendation 1: The Office of the Undersecretary of Defense for Acquisition, Technology, and Logistics and the Office of the Director of Operational Test and Evaluation of the U.S. Department of Defense and their service equivalents should acquire expertise

and appropriate tools related to model-based approaches for the requirements setting process and for test case and scenario generation for validation.

DESIGN AND DEVELOPMENT

Technological Maturity and Assessment

Conclusion 4: The maturity of technologies at the initiation of an acquisition program is a critical determinant of the program's success as measured by cost, schedule, and performance. The U.S. Department of Defense (DOD) continues to be plagued by problems caused by the insertion of immature technology into the critical path of major programs. Since there are DOD directives that are intended to ensure technological readiness, the problem appears to be caused by lack of strict enforcement of existing procedures.

Technological immaturity is known to be a primary cause of schedule slippage and cost growth in DOD program acquisition. Many studies, including those of the National Research Council (2011), the Defense Science Board (1990), and the U.S. General Accounting Office (1992) and its successor, the U.S. Government Accountability Office (2004), have discussed the dangers associated with inserting insufficiently mature technologies in the critical path of DOD design and development.

Recommendation 2: The Under Secretary for Acquisition, Technology, and Logistics of the U.S. Department of Defense (USD-AT&L) should require that all technologies to be included in a formal acquisition program have sufficient technological maturity, consistent with TRL (technology readiness level) 7, before the acquisition program is approved at Milestone B (or earlier) or before the technology is inserted in a later increment if evolutionary acquisition procedures are being used. In addition, the USD-AT&L or the service acquisition executive should request the Director of Defense Research and Engineering (the DOD's most senior technologist) to certify or refuse to certify sufficient technological maturity before a Milestone B decision is made. The acquisition executive should also

- review the analysis of alternatives assessment of technological risk and maturity;
- obtain an independent evaluation of that assessment as required in DOD instruction (DODI) 5000.02; and

- ensure, during developmental test and evaluation, that the materiel developer shall assess technical progress and maturity against critical technical parameters that are documented in the Test and Evaluation Master Plan (TEMP).

A substantial part of the above recommendation is currently required by law or by DOD instructions. Moreover, earlier NRC reports have also made similar recommendations. DOD has been moving in the wrong direction regarding the enforcement of an important and reasonable policy as stated in DODI 5000.02.

Conclusion 5: The performance of a defense system early in development is often not rigorously assessed, and in some cases the results of assessments are ignored; this is especially so for suitability assessments. This lack of rigorous assessment occurs in the generation of system requirements; in the timing of the delivery of prototype components, subsystems, and systems from the developer to the government for developmental testing; and in the delivery of production-representative system prototypes for operational testing. As a result, throughout early development, systems are allowed to advance to later stages of development when substantial design problems remain. Instead, there should be clear-cut decision making during milestones based on the application of objective metrics. Adequate metrics do exist (e.g., contractual design specifications, key performance parameters, reliability criteria, critical operational issues). However, the primary problem appears to be a lack of enforcement.

Defense systems should not pass milestones unless there is objective quantitative evidence that major design thresholds, key performance parameters, and reliability criteria have been met or can be achieved with minor product improvements.

Staged Development

Conclusion 6: There are substantial benefits to the use of staged development, with multiple releases, of large complex systems, especially in the case of software systems and software-intensive systems. Staged development allows for feedback from customers that can be used to guide subsequent releases.

The "agile development" process for software systems (discussed at the workshop) is a disciplined framework that ensures that best practices are consistently used throughout system development. A staged development appears to be natural for large-scale complex software systems, and it may also be appropriate for some hardware systems. Each of the stages must retain the functionality of its predecessor systems, at the very least to satisfy the natural expectations of the customer over time.

TESTING METHODS

The panel supports the recommendations on testing that have appeared in previous reports on this topic by the NRC. These recommendations have addressed the following issues:

- the importance of comprehensive test planning (National Research Council, 1998)
- the benefits from use of state-of-the-art experimental design principles and practices (National Research Council, 1998)
- the potential benefits from combining information for operational assessment (National Research Council, 1998)
- that testing should be carried out with an operational perspective (National Research Council, 2006)
- that testing should give greater emphasis to suitability (National Research Council, 1998)
- the benefits from the use of accelerated reliability testing methods (National Research Council, 1998)

COMMUNICATION, RESOURCES, AND INFRASTRUCTURE

Conclusion 1 highlights the need for early and clear communications about requirements. In addition, industry representatives at the workshop stressed the importance of collaboration and communication among customers and program developers, as well as participants across all aspects of system development and testing to avoid long, costly, and unsuccessful product development programs. Leading industrial companies have established programs to promote higher levels of collaboration among suppliers, manufacturers, customers, service organizations, and the ultimate users of the product.

A Data Archive

Conclusion 7: A data archive with information on developmental and operational test and field data will provide a common framework for discussions on requirements and priorities for development. In addition, it can be used to expedite the identification of

and correction of design flaws. Given the expenses and complexity of developing such an archive, it is important that the benefits of a data archive be adequately demonstrated to support development.

The collection and analysis of data on test and field performance, including warranty data, is a standard feature in commercial industries. The development of a data archive has been discussed in previous NRC reports, and we repeat its importance here. One possible reason for DOD's failure to establish a data archive is the lack of an incentive to support this and any other central activity. DOD needs to be convinced of the advantages of building and maintaining such a database and then to commission an appropriate group of people with experience in program development to develop a concrete proposal on how the data archive should be structured.

> Recommendation 3: The U.S. Department of Defense should create a defense system data archive containing developmental test, operational test, and field performance data from both contractors and the government. Such an archive would achieve several important objectives in the development of defense systems:
>
> - substantially increase DOD's ability to produce more feasible requirements,
> - support early detection of system defects,
> - improve developmental and operational test design, and
> - improve modeling and simulation through better model validation.

As DOD initiates plans to begin creation of a defense system data archive, at least three issues need immediate resolution: (1) whether the archive should be DOD-wide or should be stratified by type of system to limit its size, (2) what data are to be included and how the data elements should be represented to facilitate linkages of related systems, and (3) what data-based management structure is used. A flexible architecture should be used so that if the archive is initially limited to a subset of the data sources recommended here due to budgetary considerations, the archive can be readily expanded over time to include the remaining sources.

Feedback Loops

> Conclusion 8: Feedback loops can significantly improve system development by improving both developmental and operational test design and the use of modeling and simulation. Feedback

systems can function similarly to warranty management systems that have proved essential to the automotive industry. To develop feasible requirements, understanding how components installed in related systems have performed when fielded is extremely useful in understanding their limitations for possibly more stressful use in a proposed system. To support such feedback loops, data on field performance, test data, and results from modeling and simulation must be easily accessible, which highlights the necessity for a test and field data archive.

Field performance data are the ultimate indicators of how well a system is functioning in operational conditions. By field performance data, we also mean data on all the circumstances that can have an impact on the quality of the components, subsystems, and systems. These data include all relevant pre- and postdeployment activities, including transportation, maintenance, implementation, and storage. They could also include training data, if such data were collected objectively. Such information can and should be used to better understand the strengths and weaknesses of newly fielded systems in undertaking various missions, including such tactical information as identifying the scenarios in which the current system should and should not be used. Unfortunately, these data are rarely archived in a way that facilitates analysis.

> **Recommendation 4: After a test and field data archive has been established, the Under Secretary of Defense for Acquisition, Technology, and Logistics (USD-AT&L) and the acquisition executives in the military services should lead a U.S. Department of Defense (DOD) effort to develop feedback loops on improving fielded systems and on better understanding tactics of use of fielded systems. The DOD acquisition and testing communities should also learn to use feedback loops to improve the process of system development, to improve developmental and operational test schemes, and to improve any modeling and simulation used to assess operational performance.**

SYSTEMS ENGINEERING EXPERTISE

Conclusion 9: In recent years, the U.S. Department of Defense has lost much of its expertise in all the key areas of systems engineering. It is important to regain in-house capability in areas relating to the design, development and operation of major systems and

subsystems. One such area is expertise in the model-based design tools as discussed earlier.

Commercial companies place a great deal of importance on systems engineering expertise. This is key for system development as well as for requirements setting, model development, and testing. Unfortunately, DOD's expertise in systems engineering was decimated by congressionally mandated manpower reductions in the late 1990s and additional reductions by the services. DOD has recognized this problem and is taking steps to rectify it. However, given the time it will take to build up that expertise in house, the DOD should examine the short-term use of contractors, academics, employees of national laboratories, and others.

MANAGEMENT ISSUES

Enforcement of DOD Directives and Procedures

Conclusion 10: Many of the critical problems in the U.S. Department of Defense acquisition system can be attributed to the lack of enforcement of existing directives and procedures rather than to deficiencies in them or the need for new ones.

As workshop participants noted, there are many studies, documents, and DOD procedures relating to best practices. The problem is that they are not systematically followed in practice.

Role of a DOD Program Manager

The role of program manager is noticeably different in industry than in DOD. In industry, the program manager's tenure covers the entire product realization process, from planning, design, development, and manufacturing to even initial phases of sales and field support, and the program manager is fully responsible and accountable for all of these activities. This tenure ensures a smooth transition across the different phases of acquisition, as well as transfer of knowledge. In contrast, in DOD the tenure of a program manager rarely covers more than one phase of a project, and there is little accountability. Moreover, there is little incentive for a DOD program manager to take a comprehensive approach to seek and discover system defects or design flaws.

Recommendation 5: The Under Secretary of Defense for Acquisition, Technology, and Logistics should provide for an indepen-

dent evaluation of the progress of ACAT I systems in development when there is a change in program manager. This evaluation should include a review by the Office of the Secretary of Defense (OSD), complemented by independent scientific expertise as needed, to address outstanding technical manufacturing and capability issues, to assess the progress of a defense system under the previous program manager, and to ensure that the new program manager is fully informed of and calibrated to present and likely future OSD concerns.

Clearly, there are many details and challenges associated with developing and implementing this recommendation that are beyond the panel's scope and expertise. However, we emphasize that there are systemic problems with the current system of program management and that they are serious obstacles to the implementation of efficient practices.

1

Introduction

SCOPE OF THE STUDY

Over the past decade and a half, the National Research Council, through its Committee on National Statistics, has carried out a number of studies on the application of statistical methods to improve the testing and development of defense systems. These studies were intended to provide advice to the U.S. Department of Defense (DOD), which sponsored these studies. Unlike the earlier ones, the goal of this study was to identify current *engineering practices* that have proved successful in industrial applications for system development and testing.

The Panel on Industrial Methods for the Effective Test and Development of Defense Systems was given the following charge:

> An ad hoc committee, under the auspices of the Committee on National Statistics and the Board on Army Science and Technology, will plan and conduct a workshop that will explore ways in which developmental and operational testing, modeling and simulation, and related techniques can improve the development and performance of defense systems. The workshop will feature invited presentations and discussion to identify specific techniques that have been shown to be effective in industrial applications and are likely to be useful in defense system development.

In addition to the broad issues in its charge, the panel identified three specific topics for its focus, which we selected from a larger number that were immediately motivated by the panel's charge: finding failure modes earlier, technological maturity, and use of all relevant information for operational assessments. Our view was that these specific topics were more important and likely to benefit from greater examination.

Finding Failure Modes Earlier It is well known that an effective way to reduce costs and development times is to identify failure modes and design flaws as early as possible during the development of defense systems. What techniques are used in industry to accomplish this? Are there some generally applicable principles and practices that could be learned from the commercial sector and applied to DOD? How useful is it to test under conditions of operational realism early in system development? What aspects of the operational environment can be safely simulated and what can be ignored? What is meant by the envelope of operational performance for a system, and how far beyond that envelope should one test to discover design flaws and system limitations? Related to this, how are accelerated life tests utilized in industry? What are the advantages and disadvantages?

Technological Maturity The inclusion of hardware and software components that are not mature is often the cause of delays in defense system development and reduced performance when fielded. It is insufficient to assess the suitability and effectiveness of individual components of defense systems with respect to component-level requirements and specifications, disregarding how a component functions as part of the whole system. Such an approach represents an assessment of technological maturity in isolation, ignoring the likely environments of operational use, the impact of the employment of typical users, and other potential difficulties involving interoperability with the remaining system. A second related issue is how much of the testing resources should be allocated to just the components and how much should be devoted to testing them as part of the parent system. How do these issues differ for hardware versus software systems?

Use of All Relevant Information for Operational Assessment Data from many different sources are used to design tests and assess operational system performance. These include developmental testing, operational testing, modeling and simulation, and the same types of data from earlier stages of development for both the current system (when evolutionary acquisition is used) and for closely related systems. In evolutionary acquisition, there are also field performance data that are often available from the fielding of earlier versions of the system. As a result, information may be available from the operation of a system in very different contexts and can also involve appreciably different systems, given that the system in question will change during development. It is therefore a challenge to incorporate all of these sources of data to guide developmental and operational test design and to improve operational evaluation. Field performance data represent a particularly valuable resource since they can be extremely useful in supporting three types of feedback loops: (1) improving system design based on deficiencies experienced in the field (recognizing that field performance data can be severely incomplete), (2) improving developmental and operational test strategy by observing what system design flaws were missed in developmental and operational testing that later appeared in the field, and (3) using field performance data to validate modeling and simulation.

THE PANEL'S APPROACH

The main information-gathering activity for the study was a one-and-a-half-day workshop (see Appendix A for the program and list of speakers). The workshop was preceded by a preliminary meeting of the panel to plan the workshop, and it was immediately followed by a second panel meeting to develop the general outline of the report and some of its conclusions. There were two subsequent meetings at which panel members worked on drafts of the report.

The panel stresses that it could not, and did not, carry out a comprehensive literature review or examination of industrial engineering methods for systems development. Further, while our intentions were to address the three motivating questions relatively completely, many of the issues posed as part of the three motivating questions were not addressed by speakers at the workshop. What the report does do is highlight important techniques that have been found to be very useful in commercial industries and discusses their application in the DOD environment. These include processes for setting requirements, systems design, and testing. It was also necessary to consider the broader DOD acquisitions environment, since characteristics of that environment affect the applicability of industrial practices to DOD. Thus, the study considered availability and access to data (especially test data), availability of engineering and modeling expertise, and organizational structure of defense acquisition. The traditional issues in modeling and simulation were not covered in the workshop, except for the use of model-based design tools for requirements setting and test generation.

The panel recognizes that many, perhaps even most, of the leading-edge industrial practices discussed in this report may have been (or are currently being) used in DOD. Thus, the findings and recommendations in the report will not come as a surprise to some readers. However, the environment in DOD is very heterogeneous, and industrial best practices are currently not being followed consistently. Thus, one of the major goals of this report is to emphasize the benefits of such techniques and promote them so that their use becomes routine and is institutionalized.

The panel is also cognizant of the differences in the environment and incentive structures under which DOD operates compared with those in commercial industries. We have tried to keep these differences in mind in our analyses, findings, and recommendations. The panel believes that there are important gains to be achieved from using these industrial practices and processes.

STRUCTURE OF THE REPORT

The remainder of this report is organized as follows. Chapter 2 provides a summary of the workshop presentations and suggestions on hardware and software development processes. The following five chapters focus on the applicability of industrial practices in the DOD environment and offers the panel's conclusions and recommendations. Chapter 3 covers requirements setting; Chapter 4 covers system design and development; Chapter 5 covers testing methods; Chapter 6 covers communication, resources, and infrastructure; and Chapter 7 covers organization structure and related issues.

The agenda for the panel's workshop is provided in Appendix A. A brief overview of the defense acquisitions process is in Appendix B. Biographical sketches of panel members and staff are in Appendix C.

2

Workshop Summary

The goal of the workshop was to have presentations on leading-edge industrial practices from speakers who are (or have recently been) involved in systems development in the commercial sector. There were four primary speakers, two on software and two on hardware. The speakers were selected on the basis of their direct involvement with requirements setting, systems design and development, and system testing.

The speakers had been asked to present an overview of approaches to system development, with an emphasis on addressing the three motivating questions for the panel's work (see Chapter 1). Each set of presentations (on software and on hardware) was followed by two discussants, one with a defense perspective and one from the panel, and then general discussion.

SOFTWARE

HP-UX Continuous Development, Integration, and Test: An Agile Process[1]

The first presentation on software was by Donald Bollinger, a distinguished technologist in the Mission-Critical Business Systems Division of Hewlett-Packard (HP). He has designed and overseen the development,

---

[1]The presentation slides are available at http://www7.nationalacademies.org/cnstat/Presentations%20Main%20Page.html [November 2011].

integration, and testing of HP-UX, the operating system environment for HP's critical computer systems.

Bollinger focused on the HP-UX system and used it as a case study to describe leading-edge software development practices at HP. He noted that it is an example of an "agile" development process.[2] HP-UX is a large, complex software system with tens of millions of lines of code. It is used in mission/business-critical environments, and it is essential that very high quality is maintained release after release. It has been upgraded repeatedly, piece by piece, over the past 25 years. It has spanned four hardware architectures and dozens of platforms.

Over the past 10 years, Bollinger noted, HP moved from a "waterfall" software development process to an agile development process. (Briefly, a waterfall program proceeds from concept, to requirements, design, prototype, construction, acceptance test and final delivery. Complete and detailed requirements are emphasized. Deviations from the initial requirements are expensive and disruptive. In contrast, an Agile program begins with the same concept, and executes multiple iterations. Each iteration is a full pass from requirements to acceptance test. The first iteration will quickly [e.g., in one-tenth the time] produce an extremely minimal version of the concept. Subsequent iterations add or improve the capabilities of the product until a useful version is created, and keep going after that to create ever more useful versions. An Agile program embraces changing requirements, exploits knowledge gained in early iterations to improve designs and implementation, and encourages user feedback to guide later iterations.) One key difference between HP-UX and many DOD software systems is that HP-UX is the same basic system—only new functionalities and capabilities are added over time. The capabilities never degrade, and the customers do not change much over time.

Bollinger touched on a number of the system's key features, with an emphasis on incremental delivery of working software. There has been a substantial improvement in quality (in terms of customer defect rates as well as productivity release time) after HP switched to the continuous development, integration, and test process. Bollinger noted that HP continuously develops, integrates, and tests all elements of HP-UX to ship release criteria. Furthermore, the company starts the next release, at full throttle, the day after the last one is finished. He also emphasized the importance of not breaking legacy and of fixing defects before adding new code.

Bollinger strongly emphasized the importance of communication and collaboration with the customer and all other members of the develop-

---

[2]For the principles of the "agile manifesto," see http://agilemanifesto.org/principles.html [August 2011].

ment and testing team. Those discussions cover a variety of issues, including which requirements are "must-haves" and which are flexible, which requirements are unattainable, how specifications in the written documents should be interpreted, and information feedback from the field. Bollinger also mentioned the concept of "open development" in which the development teams share the results they have (subject to some appropriate protection). This last point may be more relevant for the contractor than for DOD personnel during developmental testing and operational testing. Bollinger also repeatedly emphasized the importance of accountability, efficiency, and cost performance in the commercial environment.

Testing in an Evolutionary Acquisition Environment: Agile with Discipline

The second presentation on software was by Sham Vaidya, an IBM distinguished engineer and the service area leader for emerging technology and architecture for IBM Global Business Services and a member of the IBM Academy of Technology. His experience is in information technology with a focus on enterprise architecture, component business modeling, business architecture, application integration, and business-oriented architectures.[3]

Vaidya discussed three case studies: a large global warranty management system for an automobile manufacturer, verification and validation of the power PC microprocessor chips in the pSeries boxes, and setting up a testing center of excellence for wireless operations for a large telecommunications client. He noted that he is a proponent of the agile software development process, and a number of his points were in common with Bollinger's.

The warranty management system was a large and complex program with about 300 million claims, an additional 16 million new claims a year, 2,000 users globally, and about 200 users interacting with the system at any given time. There were multiple data sources: faxes, batch inputs, the Web, and some defined user interface. This is somewhat similar to the types of data sources from the field in DOD applications. The major lesson IBM learned from this project was that the system could not be developed and released in "one shot"; rather a multirelease approach was necessary. (This approach had also been mentioned as a core element of agile development.) In this application, it was not possible to anticipate all of the requirements up front. New design components were added in subsequent releases,

---

[3]The presentation slides are available at http://www7.nationalacademies.org/cnstat/Presentations%20Main%20Page.html [November 2011].

based on lessons learned and feedback from the field. The effort to acquire and prepare the correct test data was, in itself, a huge project.

The power PC chip verification project dealt with the generation of test suites. The issues here are related to those in software testing. Vaidya stressed the importance of the use of a hierarchical verification approach, starting with small components and integrating more and more until the full system level is reached. The focus of the last project (testing center for wireless) was the role of various testing functions to maximize test efficiencies and ensure the timely production of high-quality software.

Discussion

One aspect of agile development, recommended by both speakers, created some controversy. This was "progressive or changing requirements" which could come even late in development. The agile manifesto notes that agile processes harness change for the developer's competitive advantage. This point met with some resistance at the workshop. Several participants raised serious reservations about using such a process in DOD's environment, in which there are already many opportunities and incentives for gaming the system. In addition, fluid requirements may lead to costly changes in system architecture (especially with hardware), introduction of new defects, and delays in system delivery. Clearly, there are systems that are suitable for staged development and multiple releases (such as those acquired in an evolutionary manner in DOD; see National Research Council, 2006) when changes in requirements will happen over time and are guided by feedback from customers.

One panel member noted that many of the concepts that are included in the agile manifesto are, by themselves, not new. It appears that, like many quality management paradigms, what is new is the disciplined environment that is promoted in the agile development process. By "disciplined environment," we refer to a systematic approach to process development that is based on accepted quality management and systems engineering principles. For example, the agile software development process is based on the 12 principles outlined in the agile manifesto (Beck et al., 2001). It emphasizes, among other things, customer satisfaction by rapid delivery of useful software, working software as the principal measure of progress, close, daily cooperation between the business people and the software developers, and sustainable development (the ability to maintain a constant pace).

HARDWARE

Effective Development and Validation Processes

The first presentation on hardware was by Jeffrey Zyburt. Now a private consultant, he spent 30 years at various positions in Chrysler, including as director of vehicle development and director of proving grounds and durability testing labs, with extensive experience in hardware development and manufacturing processes.[4]

Zyburt's presentation focused on the causes of ineffective and effective development processes, based on his experience for vehicles in the automotive industry. He first listed some of the reasons for ineffective product development:

- lack of a "dedicated" program lead and cross-functional core development team from concept to postproduction,
- ever-changing program targets and functional objectives,
- late design changes,
- no prioritization of customer requirements and no distinction between "must-haves" and "wish list,"
- late component/system supplier sourcing,
- inadequate supplier capabilities (design/development and analysis/ testing),
- no agreed on pass/fail test criteria, and
- advanced engineering and concept design and redesign that occurs along the critical path of the program timeline.

In contrast, Zyburt then listed the characteristics that are an integral part of an effective vehicle development program:

- "dedicated" upfront resources, including the program lead and a core development team, both of which are responsible and accountable until postlaunch;
- a team that is multidisciplinary (different aspects of the vehicle development) and that ensures all of the functional attributes of the vehicle can meet the program targets;
- a prioritized list of customer requirements and an identification of the sacred few or "must-haves" (based on compelling questions early in the program) and ensures that the "wish list" is well aligned with the "must haves";

---

[4]The presentation slides are available at http://www7.nationalacademies.org/cnstat/ Presentations%20Main%20Page.html [November 2011].

- defined/nonfluid functional objectives for the program;
- offline (outside the program timeline) advanced technology development;
- coordinated releases of subsystems from all disciplines so that the vehicle can be evaluated as a system for risk assessments at each milestone;
- reassessment if any program change is proposed;
- independent third-party (internal or external) assessments at each milestone with objective "go/don't go" metrics; and
- closed-loop feedback from field/warranty data on issues found and use of gap analysis (analysis of the causes of the reasons for the differences between the performance of the current system and the stated requirements) to identify scope for improvement.

Trends in Automotive Electronics Design: Current and Future Methodologies

The second presentation on hardware was by Salim Momin. Currently with SRS Enterprises, he previously was with Freescale Semiconductors, where he managed the "virtual garage" (among other activities). The objective of this organization was to understand how Freescale's customers (tier-one suppliers to automotive companies) and their customers were developing their designs.[5]

Momin noted that automotive manufacturers are moving from being component focused to being architecture focused because the latter is the key to system integration. To enable this change, companies are increasingly adopting model-based approaches to control systems engineering and requirements setting. Model-based design is an approach to codifying (formalizing) the process of taking customer requirements and translating them into systems requirements and specifications. In some cases the executable specifications can be generated, which leads to implementation. For example, in software, C code can be generated from the models using auto-code generation tools. In other words, text-based requirements are converted into mathematical equations, and mathematical analysis and simulation, visualization, and animation techniques are then used to verify and clarify the requirements. A model-based approach has many advantages, including validation of requirements, consistency checks, and resolving ambiguities in the statement of requirements and specifications.

Momin pointed to several advantages of the use of modeling in DOD's context: (1) it specifies the actual intent of the functionality so that it is very

---

[5]The presentation slides are available at http://www7.nationalacademies.org/cnstat/ Presentations%20Main%20Page.html [November 2011].

clear and precise; (2) it is reusable if it is well documented; (3) it is executable, so it gives an unambiguous functional execution; (4) it can be used to automatically generate test suites (i.e., schemes for selecting scenarios for testing the system); and (5) perhaps most importantly, the model captures knowledge that is preserved and institutionalized. In other words, it provides a formal and rigorous framework for the requirements generation process. In some cases—such as software or logic design for integrated circuits—the models can be used for implementation of the design.

Momin also mentioned that the default standard for functional modeling in the automotive industry is based on tools from MathWorks® (Stateflow®, Simulink®, MATLAB®). Other tools, such as UML and SysML, are also being used in other application areas, such as aerospace by companies like Boeing, enabled by tools from IBM. He noted other examples: General Motors uses UML for modeling and code generation of software for electronic control units used to control comfort and convenience functions of cars; and Ford uses Stateflow® and Embedded Coder® for their body electronic control units. Most engine control software is modeled using Simulink/Stateflow® and C code is auto-generated—companies doing this are General Motors, Ford, Chrysler, Toyota, and Nissan.

Adequate documentation is critical for a model-based design approach to work. Momin acknowledged the difficulty of getting engineers to spend time on documentation. He noted the availability of software tools, such as those developed by MathWorks®, which facilitate the process of documentation.

Discussion

A participant from DOD noted that the model-based design tools described by Momin are beginning to be used by defense contractors for complex systems. However, DOD itself may have limited capability in exercising these models during the review process, which is a serious limitation in collaborating with contractors.

Both Bollinger (software) and Zyburt (hardware) emphasized the importance of asking customers to prioritize their requirements into two groups: a list of "must-haves" and a "wish list." This approach has obvious advantages as it forces the customer to think carefully through the requirements at the beginning of the development process and to make tough decisions. Also, having a prioritized wish list provides considerable flexibility in trading off these requirements during design and development stages.

Zyburt repeated his point that late design changes are one of the features of an ineffective vehicle development process. Changes to system design and architecture often result in substantial cost increases, delays in development, possible introduction of additional defects, and degraded quality. This perspective conflicts with the emphasis on changing requirements in the agile manifesto, which was referred to approvingly by both speakers on software systems, Bollinger and Vaidya. It is possible that large and complex software systems are well suited for staged development and multiple releases, when the requirements over stages can change depending on feedback from customers and the field. The importance of appropriate oversight and accountability in approving design changes was also discussed by workshop participants. In feedback received after the workshop, Momin and Zyburt noted the advantages if DOD were to establish and enforce processes for evaluating the impact of changes in requirements on system design and also establish clear guidelines and criteria for accepting changes in requirements after the freeze. However, there are already guidelines and criteria in place within DOD for approving changes in requirements and design. Nevertheless, programs continue to be plagued by the occurrence of "requirements creep," suggesting that the procedures are not being followed or enforced.

Extensive communication and collaboration among the design, development, and testing teams were stressed as integral parts of leading-edge practices in the commercial sector. Another common discussion issue was ensuring maturity of new technologies since innovating on a schedule is often not possible. (This topic has been discussed in previous National Research Council reports [e.g., 2006] and unspecified DOD studies; see also discussion earlier in this chapter and in Chapter 3.) Some of the participants from industry suggested that the real problem might be lack of adherence to criteria in the assessment of new technology readiness and that there may be poor risk assessment of the impact of technology insertion and integration on systems. They speculated that this might be part of a general lack of adequate enforcement and oversight by domain experts at key milestone deliverables.

Several other issues were emphasized by more than one speaker at the workshop:

- the importance of accountability and continuity of the project management team;
- better managing the hand-off process during system development and testing so that useful information available to the developer is also available to testers;
- making clear-cut decisions during milestones—that is, "red" and "green" decisions based on objective metrics and not "yellow" ones;
- the importance of not breaking legacy and fixing defects before new components or subsystems are added; and

- the substantial benefits in using feedback loops for system improvement and for test and model improvement.

Some of the industry speakers noted at the end of the workshop that although there seem to be reasonable rules and procedures in place within DOD, it appears they are not properly implemented by appropriate checks and balances. They speculated that this is probably the major hindrance to the improvement of defense acquisition. In fact, one of the speakers from industry noted: "The good news is, all the studies you [have done] . . . you know 80 percent [of the best practices and guidelines needed] is already there. All you've got to do is go out and do what you wrote down, and you'll be in great shape."

3

Requirements Setting

In this and the next four chapters, the panel assesses the industry practices described at the workshop and discusses their applicability within defense acquisitions. As noted in Chapter 1, a number of the suggestions made at the workshop are already represented in the documents specifying the acquisition policies and procedures of the U.S. Department of Defense (DOD); practices that DOD has been trying to implement; or practices that previous National Research Council (NRC) panels or other advisory bodies have recommended. For these situations, we have chosen not to make new policy or procedural recommendations. In cases for which it appears those practices are not being followed widely, we have reiterated the benefits and the need to widely adopt and institutionalize the practices. In other cases, we offer additional arguments for following the previously recommended procedures. Our recommendations are restricted to situations in which the panel believes that the practices are either new, have elements that are new, or in which DOD practices are moving in the opposite direction.

In this chapter, we consider requirements setting in light of the practices discussed at the workshop. The panel recognizes that requirements are often initially set at overly optimistic levels so that a program will attract funding. This issue is beyond the scope of our study and is not explicitly addressed here.

COMMUNICATION WITH USERS

Conclusion 1: It is critical that there is early and clear communications and collaboration with users about requirements. In particular, it is extremely beneficial to get users, developers, and testers to collaborate on initial estimates of feasibility and for users to then categorize their requirements into a list of "must-haves" and a "wish list" with some prioritization that can be used to trade off at later stages of system development if necessary.

This conclusion reflects the need for continuous exchange and involvement of users in the development of requirements. User input can assist in assessing cost and mission effectiveness of a design and can aid in the development of the "analysis of alternatives."[1] Although continuous involvement in the development of requirements by users does occur in DOD—for example, the Army designates a capabilities manager for the U.S. Army Training and Doctrine Command to represent the user on a program—it does not appear to be emphasized as much or conducted as extensively as in industry.

The industry practice of asking customers to separate their needs into a list of "must-haves" and a "wish list" is especially appealing. It imposes discipline on customers: they are forced to carefully examine a system's needs and capabilities and any discrepancies between them and, thus, make decisions early in the development process. Communication and collaboration also ensure that all parties, including the user, the program manager, the developer, and the tester, agree on the required performance levels of a system. Although elements of this concept have been implemented in DOD through the use of threshold and objective levels for requirements and by banding requirements and key system attributes, with appropriately higher authority approval required for any change, we emphasize that more can be done for more effective requirements setting.

---

[1]An analysis of alternatives (AoA) is part of several steps in the Joint Capabilities Integration and Development System (JCIDS), which assesses cost and mission effectiveness, given levels of performance and suitability. In JCIDS (a formal DOD procedure that defines requirements and evaluation criteria for defense systems in development) a required capability (e.g., defeat an Integrated Air Defense System) is evaluated through a capability-based analysis (CBA) and then by an AoA to develop system attributes as a function of required levels of performance and suitability. However, only system attributes are provided as "requirements" to the development and test community. Currently, there is no quantitative way to assess the impact of not meeting a system requirement on accomplishing the mission. If, on the other hand, the JCIDS/CBA/AoA process provided a quantitative linkage between mission accomplishment and system attributes, the acquisitions community would have an effective method for making decisions on threshold levels set by the requirements process and for understanding the cost effectiveness of changing those requirements.

FEASIBILITY AND COSTS

The steps proposed above must be complemented by rigorous assessment of feasibility and costs. Such an assessment will ensure that the user and the developer understand and agree that, although some additional capabilities or features may be useful add-ons, they should be sacrificed to ensure that the system attains its necessary levels of effectiveness and suitability and that they do so at an acceptable cost and in a timely manner. The panel appreciates the challenges involved in establishing shared estimates of feasibility at the outset and in making tradeoffs during requirements setting and development for major DOD acquisitions. Nevertheless, we strongly encourage the systematic approach and rigorous exchange of ideas that are part of this process.

As the workshop speakers emphasized, it is important to use input from the test and evaluation community in the setting of initial requirements. Testers can identify requirements that are either difficult or impossible to test or those that are ambiguous or are mutually inconsistent. Therefore, input from testers is a critical part of system design. In staged development, input from users and from the field can also be very informative in understanding what an early system can and cannot do.[2]

CHANGES IN REQUIREMENTS

Conclusion 2: Changes to requirements that necessitate a substantial revision of a system's architecture should be avoided as they can result in considerable cost increases, delays in development, and even the introduction of other defects.

Once a system's architecture is set, changing requirements can be extremely expensive, is likely to add considerably to development time,

---

[2]Bell (2008) strongly advocates the use of a team approach to the setting of requirements. He states that the benefits from the testers and the program management offices becoming a team from the beginning of acquisition has at least six benefits: (1) more realistic requirements, (2) verifiable requirements, (3) verifiable specifications, (4) requirements and specifications that are understood, (5) appropriate testing-related schedule, budget, and infrastructure, and (6) contractors prevented from under- or overbidding the test and evaluation part of their proposal. With a team approach in place, the system integration laboratory becomes a useful preparatory time and place. Testers are encouraged to double-check that proper reliability growth is planned and executed, to interact with independent operational testers, and to plan and execute developmental test and evaluation thoroughly enough to virtually ensure success in initial operational test and evaluation. With this approach, program management offices, with only a small initial investment, can potentially save large sums of money.

and can introduce additional failure modes and design flaws.[3,4] Having stable requirements during development allows the system architecture to be optimized for a specific set of specifications, rather than being modified in a suboptimal manner to try and accommodate various updates to the requirements. At the same time, however, there also must be some flexibility that allows for modifications that are responsive to users' needs and changing environments.

A previous NRC report (2008:50) discussed the tension between these two goals:

> One must clearly establish a complete and stable set of system-level requirements and products at Milestone A. While requirements creep is a real problem that must be addressed, some degree of requirements flexibility is also necessary as lessons involving feasibility and practicality are learned and insights are gained as technology is matured and the development subsequently proceeds. Certainly control is necessary, but not an absolute freeze. Also, planning ahead for most likely change possibilities through architectural choices should be encouraged, but deliberately managed, a concept encouraged herein.

The panel endorses this statement and notes that it is consistent with the views expressed by the participants at our workshop.

As noted above, greater fluidity in requirements may be quite reasonable (and even desirable) for software systems: reworking may be feasible later in development for software systems than for hardware systems. And even with hardware systems, changes to requirements may be relatively easy for systems that are acquired in an evolutionary manner. The key is that the process for changing requirements should be well managed, with adequate oversight, clear accountability, and enforcement of the rules. In

---

[3]Thompson (1992:738-739) notes that the F-16a fighter is good example of the effects on system reliability when one is allowed to keep changing requirements: "Instead of the simple, austere, pure fighter it was originally planned to be, the air force made it into a dual purpose aircraft, used to attack ground targets as well as a dog-fighter. This increased its price 75 percent and increased its weight from ten tons to over twelve, with a proportional reduction in acceleration. It also increased the plane's complexity, owing to the installation of additional avionics, radar, and electronic countermeasures, with proportional reductions in reliability and maintainability."

[4]Tangentially, we note that it often makes little difference whether the ultimate system passes or slightly fails achievement of the requirements in the fielded system. For example, compare the situations in which a jet fighter in development either flies at better than Mach 2 or flies at only Mach 1.8. That is unlikely to make an important difference as to the successful completion of missions. Instead, what is important is that once the system is fielded, the user needs to have a comprehensive understanding of precisely what the system can and cannot do. That is why it is very important to test to failure in development whenever possible rather than test exclusively to requirements.

particular, input from engineers as to the feasibility of any changes to requirements needs to play a key role in decisions as to whether or not to permit any requested changes.

The panel recognizes that existing DOD regulations mandate that changes in requirements go through a rigorous engineering assessment before they are approved. However, it appears that these regulations are not being followed: there are many instances in which requirements continue to change throughout development, including reductions that result from concerns about feasibility.

USE OF MODEL-BASED DESIGN TOOLS

Conclusion 3: Model-based design tools are very useful in providing a systematic and rigorous approach to requirements setting. There are also benefits from applying them during the test generation stage. These tools are increasingly gaining attention in industry, including among defense contractors. Providing a common representation of the system under development will also enhance interactions with defense contractors.

Modeling and simulation tools are used widely in DOD, but use of the term "model-based design" here is narrower. The focus is on the use of tools to formally translate and quantify requirements from high-level system and subsystem specifications, assess the feasibility of proposed requirements, and help examine the implications of trading off various performance capabilities (including various aspects of effectiveness and suitability, such as durability and maintainability). A recent presentation by the National Defense Industrial Association (NDIA) engineering division's modeling and simulation committee (2011) refers to this as model-based engineering (MBE) and defines it as "an approach to engineering that uses models as an integral part of the technical baseline that includes the requirements, analysis, design, implementation, and verification of a capability, system, and/or product throughout the acquisition life cycle" (p. 7). The NDIA report notes that MBE can also include the use of physics-based models, but these are not part of the discussion here.

These tools start at a high level, when the key performance parameters or the high-level requirements are first specified. System-level requirements then flow down to subsystem- and component-level requirements, following the classic V-diagram of systems engineering. The process allocates the high-level requirements to a more detailed functional design and functional architecture for various component systems. As this happens, the model becomes more refined and acquires higher fidelity.

As described at the workshop, this approach has many benefits:

- It provides a formal specification of the actual intent of the functionality so that it is very clear and precise.
- It is reusable if it is well documented.
- It is executable, so any ambiguities can be identified.
- It can be used to automatically generate test suites.
- Perhaps most importantly, the model captures knowledge that can be preserved and institutionalized.

This is a very good way to have a formal understanding of the specification (need) and performance (deliverable) of the intended system. This approach is now used in some programs; it needs to be expanded and needs to include supplier performance models.

The model-based approach also provides a platform for common and consistent use of terminology and codification of requirements. This consistency supports the greater acceptance of performance characteristics by the contractor, program manager, users, and testers. It allows for validation or refinement of requirements by domain experts. Such models can also be used in simulation environments to assess technology readiness. They can also allow for the linkage of system-level performance requirements to the performance of sub-systems and components.

Furthermore, an overall modeling and simulation-based vision is crucial for identifying where initial efforts should be concentrated to achieve the required performance levels. Then, as development proceeds, modeling and simulation can be used to ensure that subsequent efforts remain focused on what is needed to achieve performance goals. Such a comprehensive approach to the modeling of system performance can and should be used as a repository of information on system performance, initially fed by engineering knowledge gained from previous systems and then informed and updated by test data. The modeling tools also facilitate system testing, integration, and automated code generation for specific tasks. In addition, they provide a convenient framework to archive relevant information on all past tests. There are also modeling tools specifically designed to check integration issues. The architecture at the higher levels is the integration platform. Without such tools the integration of the full system is likely to be problematic.

We note, however, that the extent to which legacy models for related systems are used for this purpose will depend on the system in question and how related the new system is to previous systems. Even relatively modest changes in a system may make legacy models and simulations poor representations of stresses and strains, etc., and, as a result, any legacy modeling and simulation needs to be rigorously validated for use on a new system.

Industries are increasingly using such model-based design tools to

---

BOX 3-1
General Motors Develops Two-Mode Hybrid Powertrain with
Model-Based Design Reduces Development Time by 2 Years
with Math and Simulation-Based Tools from MathWorks®
[Press release]

General Motors Company (GM) has developed its Two-Mode Hybrid power-train control system using Model-Based Design. By using math and simulation-based tools from The MathWorks®, GM designed the powertrain prototype within 9 months, shaving 24 months off the expected development time. . . . By adopting Model-Based Design, where the development process centers around a system model, GM engineers increase time savings. Also, by verifying the control system before hardware prototyping and by using production code generated from the controller models, GM has rolled out production vehicles featuring the hybrid powertrain within four years of starting the control system design process. The ability to reuse design information has helped the global development teams foster more efficient communication and reduced response time, eliminating integration issues. . . . GM used MATLAB®, Simulink®, and Stateflow® to design the control system architecture and model all the control and diagnostic functions. Real-Time Workshop Embedded Coder provided the capability to generate production code from the models, and Real-Time Workshop and hardware-in-the-loop (HIL) simulators helped verify the control system.

SOURCE: MathWorks® (2009). Available: http://www.mathworks.com/company/pressroom/General-Motors-Develops-Two-Mode-Hybrid-Powertrain-With-Model-Based-Design.html [November 2011]. Reprinted with permission.

---

assess the feasibility of requirements. In some cases, the entire architecture of complicated systems is driven by modeling tools such as those employed by General Motors: see Box 3-1.[5] Some DOD programs are obviously far more complex than automotive programs and so can benefit greatly from these tools. The NDIA report suggests that defense contractors are already using these tools, although the level of usage may vary considerably.

DOD should have expertise in these tools and technologies so that the agency can use them in its interactions with contractors and users. It is crucial that DOD at least actively participate, if not lead, in the development of the relevant models. Operational effectiveness models are critical for requirements setting; systems performance models are

---

[5]See http://www.mathworks.com/company/pressroom/General-Motors-Develops-Two-Mode-Hybrid-Powertrain-With-Model-Based-Design.html [November 2011].

critical for assessing feasibility. The latter can serve as a critical tool for collaboration between contractors and DOD. For example, they allow for the traceability of requirements since everyone is working from the same set of assumptions, leading to a disciplined approach in the development process. The model is also a feedback mechanism, providing answers to "what if" questions about the functioning of the system. For all these reasons, DOD should not rely completely on contractors to develop and use this capability. Given their importance, performance models should be part of contract deliverables, just as computer-assisted design models are now, and their review should be a key part of any milestone decision.

> **Recommendation 1: The Office of the Under Secretary of Defense for Acquisition, Technology, and Logistics, and the Office of the Director of Operational Test and Evaluation, and their service equivalents should acquire expertise and appropriate tools related to model-based approaches for the requirements-setting process and for test case and scenario generation for validation.**

This expertise will be very beneficial in collaborating with defense contractors and in providing a systematic and rigorous framework for overseeing the entire requirements generation process. The expertise can be acquired inhouse or through consulting and contractual agreements.

4

Design and Development

This chapter considers three key aspects of industrial engineering methods for system design and development: (1) the need to assess the technological maturity of subsystems and components prior to insertion in a defense system in development, (2) the need to use objective metrics for assessment, and (3) the advantages of staged acquisition. These topics were discussed at the panel's workshop (see Appendix B).

THE IMPORTANCE OF TECHNOLOGICAL MATURITY

Consequences of Using Immature Technology

Conclusion 4: The maturity of technologies at the initiation of an acquisition program is a critical determinant of the program's success as measured by cost, schedule, and performance. The U.S. Department of Defense (DOD) continues to be plagued by problems caused by the insertion of immature technology into the critical path of major programs. Since there are DOD directives that are intended to ensure technological readiness, the problem appears to be caused by lack of strict enforcement of existing procedures.

There are many studies that describe problems caused by inserting insufficiently mature technologies in the critical path of acquisition programs for both DOD and commercial companies (see, e.g., National Research Council [2010a]): see Box 4-1. This is a primary cause of schedule

---

BOX 4-1
Use of Immature Technologies: Consequences

Four examples of conclusions from major studies of the consequences of using immature technologies are noteworthy.

1. The "Streamlining Study" of the Defense Science Board was never published, but the Institute for Defense Analysis (1991) produced IDA Paper P-2551, which covered some 100 major defense acquisition programs, reached a firm conclusion that failure to identify technical issues, as well as real costs, before entering into full-scale development—now referred to as engineering and manufacturing development—was the overwhelming cause for subsequent schedule delays and the resulting cost increases.

2. The U.S. General Accounting Office (1992:49) stated: "Successful programs have tended to pursue reasonable performance objectives and avoid the cascading effects of design instability. . . ."

3. More than a decade later, the U.S. Government Accountability Office (2004:2) found: "FCS [Future Combat System] is at significant risk for not delivering required capability within budgeted resources. Three-fourths of FCS needed technologies were still immature when the program started. The first prototypes of FCS will not be delivered until just before the production decision. Full demonstration of FCS ability to work as an overarching system will not occur until after production has begun." The report also concluded that based upon the experiences of past programs, the FCS strategy was likely to result in cost overruns and delays. In fact, the FCS program was terminated about 6 years later.

4. At a November 30, 2005, meeting of the Naval Studies Board of the National Research Council, the then newly appointed Department of the Navy acquisition executive, Dr. Delores Etter reported that she had just attended her first Defense Acquisition Board review, which was for the DDG-1000 (guided missile destroyers) program. She had anticipated that technologies for the program would be an issue with the Undersecretary of Defense (AT&L), DOD's top acquisition executive but they were not. The acquisition team had identified 10 high-risk areas that would have to mature in parallel for the acquisition program to meet its performance goals, and the program was approved for entry into engineering and manufacturing development. About 3-1/2 years later, in the summer of 2008, the Department of the Navy requested, and received approval for, termination of the prohibitively expensive program after having spent $10 billion on the first two ships.

---

slippage and cost growth in DOD program acquisition, and it often results from the overly optimistic confidence of developers in their abilities to convert technological advances into developing reliable components and subsystems and doing so in a timely manner. The terminations of the FCS (the Army's future combat system) and DDG-1000 (the Navy's Zumwalt class of guided missile destroyers) programs years after their entry into engineering and manufacturing development are strong evidence of the very adverse result of incorporating multiple immature technologies in the critical paths of large complex product developments.

The dangers of immature technology are just as critical in the commercial sector. Globalization and rapid advances in technology have put immense pressure on industry to offer "on-going" new products with the latest technological features. This pressure in turn has led to shorter product development cycles, increasing the risk of introducing immature and infeasible new technologies. Unlike the situation in DOD, product launch dates in many parts of the commercial sector, such as the automotive industry, are sacred. Any slippage in a product launch date has serious financial implications for automotive companies: they range from millions of dollars in lost revenues for every day's delay in product launch to inflicting major chaos in the entire supply chain, with a supplier who may be 10,000 miles away, to the marketing group that has already made extensive plans and commitments. And launching a product that is not fully ready also has serious cost implications, including high warranty costs and, more importantly, lost customer goodwill. Clearly, a major slip in quality at launch has severe consequences; the product may never be able to sell at planned volumes, resulting in major losses for the company.

Faced with the above challenges, top management in the commercial sector is increasingly approving "pre-spend" money for major programs. This pre-spend money is spent on conducting technical feasibility studies on perceived program challenges while the program details are still being finalized for program approval.[1] The challenges can include a wide range of activities, such as establishing feasibility of aggressive exterior styling, kicking off die development for major body panels that have long lead times, and studying the feasibility of adapting a new powertrain and getting better cost estimates on the project. The pre-spend money is often 1-2 percent of the cost of the overall program. In recent years, major industry programs have been cancelled or delayed on the basis of the results of the technical feasibility analysis conducted through pre-spend money, thereby enabling the automakers to prevent major losses later in the process.

Speakers at the workshop emphasized the importance of getting considered opinions from qualified domain experts about the adequate maturity of new technologies or about new applications of existing technologies. It was evident that the commercial sector also places a great deal

---

[1]DOD has provided analogous funding for reducing major defense acquisition program technology risks and for demonstrating the value of new technologies in separately funded "advanced technology demonstrations" and now in the new technology development phase of major defense acquisition programs.

of emphasis on not risking failure by including an unproven technology advance in a critical path of a new program.

There are indeed examples in DOD where programs have managed this issue successfully,[2] so the department has exhibited the capability to properly assess technological maturity. What is needed is a way to instill a willingness to acquire independent expert input and a collaborative spirit in those leading future programs. Such a culture is the responsibility of the most senior DOD acquisition executives and of the secretary of defense. The problems result from the different cultures and practices of the different participants in the requirements development process, the acquisition process, and the resource allocation process—not in stated DOD policies and procedures contained in DOD directives.

The *Technology Readiness Assessment Deskbook*

The current U.S. Department of Defense Instruction (DODI) 5000.02 of December 8, 2008 (which is consistent with the current DODI 5000.01 certified current as of November 20, 2007) contains the following guidance/requirement regarding technology for acquisition programs[3]:

> Technology for use in product development (starting at Milestone B) "shall have been demonstrated in a relevant environment or, *preferably, in an operational environment* [emphasis added] to be considered mature enough to use for product development. . . . If technology is not mature, the DOD component shall use alternative technology that is mature and that can meet the user's needs or engage the user in a dialog on appropriately modifying the requirements." In addition, the current 2009 *Technology Readiness Assessment (TRA) Deskbook* (p. C-5) defines "hardware" readiness levels as follows:
>
> - TRL 7 as "System prototype demonstrated in an operational environment" and
> - TRL 6 as "System/subsystem model or prototype demonstrated in a relevant environment."

---

[2]A recent report on the F-A-18E/F Super Hornet Development Program is an example of the Navy's ability to control the technological maturity in a major DOD acquisition program. As noted in the report (Center for Naval Analysis, 2009:16), the program "did not over reach on technology or capability demands." The collaboration of all the parties "allowed the E/F program to develop a clear and focused set of requirements that was simply stated and understood by all. The technology for all requirements was either already in hand, or all agreed to defer the requirement to a later block upgrade when the technology was ready" (p. 28).

[3]See DODI 5000.02 Enclosure 2, paragraph 5.d. (4). Available: http://www.dtic.mil/whs/directives/corres/pdf/500002p.pdf [August 2011].

The current (2009) *Technology Readiness Assessment (TRA) Deskbook* does not refer to the "preferred TRL 7" when describing the readiness assessment process for evaluating technology readiness for Milestone B. Rather, it is Title 10 of the U.S. Code (Section 2366b) that requires that the milestone decision authority (the person so designated for each program) certify technologies used at Milestone B have been demonstrated to perform at level TRL6.[4] This was not true in the previous version of the *TRA Deskbook,* which followed the DODI 5000.02 guidance.[5]

The current 2009 *TRA Deskbook* also describes an elaborate process for the preparation of technology readiness assessments involving a suggested schedule of 11 months and the selection of an integrated product team consisting of a balanced set of subject matter experts (SMEs) from DOD components, other government agencies, and possibly, nongovernment entities. Significant attention and space are devoted to the authorities of various parties, the "equitable processes" for selecting subject matter experts, and the desire to arrive at a single agreed-on readiness assessment. However, how to deal with different interpretations of, or opinions on, technological maturity is not a significant subject in the *Deskbook.*

The panel concludes that the philosophy behind DODI 5000.02 is adequate and that the statements about the preferred levels of technology readiness (i.e., TRL 7) for approval at Milestone B are appropriate. However, we have two concerns. One is that the guidance for implementation in the 2009 *TRA Deskbook* is not adequate (i.e., the sole focus on TRL 6 for Milestone B approval). The second is the insufficient discipline exhibited by most program managers and most DOD acquisition executives, with regard to both the technological maturity for individual components and the integration of multiple components involving interrelated technologies.

Implementation of DOD Instructions and Directives

The panel also concludes that there is a significant weakness in DOD's implementation of its own *Directive and Instruction* for acquisition pro-

---

[4]See DOD *Technology Readiness Assessment (TRA) Deskbook*, Section 1, paragraph 1.3.1. Available: http://www.dod.gov/ddre/doc/DoD_TRA_July_2009_Read_Version.pdf [August 2011].

[5]The 2003 *TRA Deskbook* stated (available: http://www.dod.mil/ddre/doc/May2005_TRA_2005_D0D.pdf [August 2011]):
- a central theme of the acquisition process is that technology employed in system development should be "mature" before system development begins (see p. ES-1);
- for Milestone B, readiness levels of at least TRL6 are typical (TRL 7 preferred); and
- for Milestone C, readiness levels of at least TRL 8 are typical (TRL 9 preferred).

grams. The proposed solution should not lower the standard in DOD's instruction to the level of just what is required by the U.S. Code (i.e., what happened in the revision of the *TRA Deskbook* from 2003 to 2009). Good industry practices as well as past successful (in contrast with unsuccessful) DOD acquisition programs support a higher level of technological readiness than has been, and is being, exhibited in most recent and current DOD acquisition programs. This view is strongly supported by the report of the first Director of Defense Research and Engineering (DDR&E) to Congress on the technical maturity and integration risk of major DOD acquisition programs.[6]

The comments from industry participants at the workshop and from several GAO reports (U.S. General Accounting Office, 1999; U.S. Government Accountability Office, 2006) indicate that, in general, technological readiness levels for commercial products are higher than they are for DOD programs. There are several possible reasons for this difference. One is that commercial products are vulnerable to product liability lawsuits and product warranties, both of which drive comprehensive performance and reliability testing prior to product introduction on the market. In contrast, with very few exceptions, DOD does not require warranties, nor is the original equipment manufacturer (the contractor) held liable for deficiencies, as are commercial manufacturers. Additionally, most DOD products are at the leading edge of technology in the hope of providing a competitive edge over potential adversaries. Notwithstanding these differences, DOD places too little attention, in general, on technological readiness prior to beginning system development.

Some of the industry participants at the panel's workshop suggested that the real problem might be the lack of adherence to criteria in the assessment of new technological readiness. In addition, it was noted that there may be poor risk assessment of the effects of technology insertion and integration on systems.

Recommendation 2: The U.S. Department of Defense (DOD) Under Secretary for Acquisition, Technology, and Logistics (USD-AT&L)

---

[6]This report (U.S. Department of Defense, 2010) was written to comply with the Weapons Systems Acquisition Reform Act of 2009 (Public Law 111-23), which requires that the DDR&E submit an annual report. The report, covering 2009, was critical of the technological readiness levels assigned to technologies in the Joint Tactical Radio System and wideband networking waveform, as well as the technological readiness levels used in the Army's first increment of its brigade combat team modernization effort. The particular programs reported on are not as important as the fact that the DDR&E's critical assessment was either not available to the relevant acquisition decision authority before Milestone B or it was available to, but not appropriately acted on by, the cognizant decision authority before or at the Milestone B decision point.

should require that all technologies to be included in a formal acquisition program have sufficient technological maturity, consistent with TRL (technology readiness level) 7, before the acquisition program is approved at Milestone B (or earlier) or before the technology is inserted in a later increment if evolutionary acquisition procedures are being used. In addition, the USD-AT&L or the service acquisition executive should request the director of defense research and engineering (DOD's most senior technologist) to certify or refuse to certify sufficient technological maturity before a Milestone B decision is made. The acquisition executive should also

- review the analysis of alternatives assessment of technological risk and maturity;
- obtain an independent evaluation of that assessment, as required in DODI 5000.02; and
- ensure, during developmental test and evaluation, that the materiel developer assesses technical progress and maturity against critical technical parameters that are documented in the test and evaluation master plan.

We are aware that a substantial part of the above recommendation is currently required by law or by DOD instructions. In particular, DODI 5000.02 obligates DDR&E to perform an independent technology readiness assessment of major defense acquisition programs prior to Milestones B and C. The director of developmental test and evaluation is supposed to provide an assessment of the test process and results to support this readiness review. Furthermore, DODI 5000.02 requires a cost assessment and program evaluation. In addition, all of the military services currently perform an operational test readiness review and must certify that the system is ready for dedicated initial operational test and evaluation. These certifications, required by DODI 5000.02, have varying degrees of depth and credibility. Recently, the USD-AT&L began performing an independent assessment of operational test readiness, and the director of developmental test and evaluation is tasked to support this effort.

But the panel believes that DOD has been moving in the wrong direction regarding enforcement of an important and reasonable policy as stated in DODI 5000.02. The recommendation of an earlier report (National Research Council, 2006) was also concerned with immature technologies. Our recommendation supports and modifies the earlier one. Our intent is to make it more difficult for advocates to incorporate immature technologies into the critical paths of major DOD acquisition programs.

USE OF OBJECTIVE METRICS FOR ASSESSMENT

Conclusion 5: The performance of a defense system early in development is often not rigorously assessed, and in some cases the results of assessments are ignored; this is especially so for suitability assessments. This lack of rigorous assessment occurs in the generation of system requirements; in the timing of the delivery of prototype components, subsystems, and systems from the developer to the government for developmental testing; and in the delivery of production-representative system prototypes for operational testing. As a result, throughout early development, systems are allowed to advance to later stages of development when substantial design problems remain. Instead, there should be clear-cut decision making during milestones based on the application of objective metrics. Adequate metrics do exist (e.g., contractual design specifications, key performance parameters, reliability criteria, critical operational issues). However, the primary problem appears to be a lack of enforcement.

There should be clear-cut decision making during milestones based on objective metrics. Adequate metrics do exist (e.g., contractual design specifications, key performance parameters, reliability criteria, critical operational issues). However, the primary problem, once again, appears to be lack of enforcement by the component and Office of the Secretary of Defense senior managers responsible for acquisition program oversight. More than one speaker at the workshop said that it is key that defense systems should not pass milestones unless there is objective, quantitative evidence that major design thresholds, key performance parameters, and reliability criteria have been met or can be achieved with minor product changes.

The lack of consistent use of objective, quantitative metrics occurs at many points during defense acquisition:

- the generation of system requirements (see Chapter 3);
- the timing of the delivery of prototype components, subsystems, and systems from the developer to the government for developmental testing;
- the delivery of production-representative system prototypes for operational testing; and
- the passage of systems into full-rate production.

The transition from developmental testing to dedicated initial operational test and evaluation (IOT&E) is often driven by schedules rather than the availability of production-representative articles with mature systems. Articles should not be delivered to IOT&E until the system is performing at a level that will meet operational requirements, as determined by a disciplined operational test readiness review, noted above. It is counterproductive to place a system into operational testing when its reliability is 20 percent or 30 percent below what is required, with the hope that enough failure modes will be discovered during operational testing to raise the reliability to the required level. More often than not, such a system will need further development and its operational testing will likely need to be repeated.

The passage of systems into full-rate production is typically justified on the basis of the results of a comprehensive operational test, which includes assessment of both effectiveness and suitability. From 2001 through 2006, DOT&E found that 15 of 28 systems undergoing IOT&E either were not operationally suitable or were suitable with limitations. Of these 28 systems, 9 were found to be either not effective or effective with significant deficiencies. However, all these systems were fielded, often with the deficiencies that had been identified during initial operational test and evaluation (see Defense Science Board, 2008).

Although the decision as to which systems in development are and are not fielded is complex, having a greater degree of rigor in decisions would reduce the chance of systems being delivered to the field and failing to meet their requirements. Such failure is particularly common with respect to system suitability. In such cases, systems often do not go back to development. Rather, a greater number of systems are purchased to ensure adequate availability—since systems may fail in the field or be under repair—thereby greatly increasing the life-cycle costs.[7]

STAGED DEVELOPMENT WITH AN EMPHASIS ON SOFTWARE

As noted in another NRC report (2010:1): "Current fielding cycles are, at best, two to three times longer than successful commercial equivalents . . . representing multiyear delays in delivering improved IT systems to warfighters and the organizations that support them. As a result, the DOD is often unable to keep pace with the rate of IT innovation in the commercial marketplace. . . ." A particular issue is the growing impor-

---

[7]Adolph (2010:53) provides an excellent discussion of these issues: "Rigorous enforcement of key requirement thresholds, along with emphasis on performance in the intended mission environment, should be the norm when entering System Development and Demonstration. Issues that need to be addressed in relation to requirements setting include technology readiness, the translation of requirements into design criteria, with attention to testability at the subsystem and system levels, as well as defining thresholds for key performance parameters."

BOX 4-2
Benefits of Agile Development

In examining current DOD processes for acquiring IT systems and comparing them with the processes adopted by leading-edge firms in the commercial sector, the committee found stark differences. DOD is hampered by a culture and acquisition-related practices that favor large programs, high-level oversight, and a very deliberate, serial approach to development and testing (the waterfall model). Programs that are expected to deliver complete, nearly perfect solutions and that take years to develop are the norm in DOD. In contrast, leading-edge commercial firms have adopted agile approaches that focus on delivering smaller increments rapidly and aggregating them over time to meet capability objectives. Moreover, DOD's process-bound, high-level oversight seems to make demands that cause developers to focus more on process than on product, and end-user participation often is too little and too late. These approaches are counter to agile acquisition practices in which the product is the primary focus, end users are engaged early and often, oversight of incremental product development is delegated to the lowest practical level, and the program management team has the flexibility to adjust the content of the increments in order to meet delivery schedules. The committee concluded that the key to resolving the chronic problems with DOD acquisition of IT systems is for DOD to adopt a fundamentally different process—one based on the lessons learned in the employment of agile management techniques in the commercial sector. Agile approaches have allowed their adopters to outstrip established industrial giants that were beset with ponderous, process-bound, industrial-age management structures. Agile approaches have succeeded because their adopters recognized the issues that contribute to risks in an IT program and changed their management structures and processes to mitigate the risks . . . for the DOD to succeed in adopting new approaches to IT acquisition, the first step is to acknowledge that simply tailoring the existing processes is not sufficient. DOD acquisition regulations do permit tailoring, but the committee found few examples of the successful application of the current acquisition regulations to IT programs, and those that were successful required herculean efforts or unique circumstances. Changes broader than tailoring are necessary; they must encompass changes to culture, redefinition of the categories of IT systems, and restructured procurement, development, and testing processes as identified in this report. In the aggregate, these changes must realign processes that today are dominated by deliberate approaches designed for the development of large, complex, hardware-dominated weapons systems to processes adapted to the very different world of software-dominated IT systems."

SOURCE: National Research Council (2010a:ix-x).

tance of software (sub)systems, and the functionality of defense systems is increasingly dependent on extremely complicated software.

> **Conclusion 6: There are substantial benefits to the use of staged development, with multiple releases, of large complex systems, especially in the case of software systems and software-intensive systems. Staged development allows for feedback from customers that can be used to guide subsequent releases.**

The workshop speakers on software systems emphasized staged development as part of "agile" development processes: see Box 4-2. In the panel's view, many elements of the agile processes are not new. What is needed, however, is a systematic approach that ensures that these practices are consistently used throughout system development. If properly implemented, these practices would ensure that defects and weaknesses in a system are detected early so that they can be addressed inexpensively.

Staged development appears to be natural for large-scale complex software systems. The use of staged development may also be appropriate for some hardware systems: two examples of situations in which substantial upgrades to fielded systems provided a substantial increase in war fighting capability are the Apache helicopter and the M-1 tank.

A good example of the applicability of agile development to hardware systems is that of the F-A-18E/F, a twin-engine carrier-based multirole fighter aircraft mentioned in footnote 3, where it was stated that the technologies were not inserted in a release until they were determined to be fully ready. This approach is consistent with the agile philosophy. However, each of the stages must retain the functionality that all the predecessor systems had, at the very least to satisfy the natural expectations of the customer over time. We note, however, that with fluid requirements, the most challenging job is to select the systems architecture in a way that can accommodate the likely changes in requirements over the several anticipated stages of development. Meeting this challenge requires foresight as to what capabilities may ultimately be requested and in designing the architecture in a way that does not make the ultimate system overly complicated, heavy, or expensive.

5

Testing Methods

Robust testing is an important part of effective system development. It can lead to early detection and correction of design deficiencies, and it facilitates high quality and reliability throughout system development. Testing at the U.S. Department of Defense (DOD) has been a subject of several previous National Research Council (NRC) reports. This section summarizes the conclusions and recommendations from those reports that are relevant to this panel's charge and offers additional analysis and suggestions.

TESTING AS A CONTINUOUS PROCESS FOR LEARNING

Operational testing and evaluation (OT&E) is intended to support a decision to pass or fail a defense system before it goes into full-scale production, but this practice has not been consistently followed by DOD. The National Research Council (1998) proposed a new paradigm recommending that testing be viewed as a continuous process of information gathering and decision making in which OT&E plays an integral role.

The new paradigm stressed the importance of adding operational realism to developmental testing. A key motivation for this focus, which is relevant to this report, is to discover design flaws much earlier in system development than currently occurs, when such defects are much less expensive to fix. It is well known that operational testing unearths many design problems missed in earlier developmental testing due to the better representation of operational realism. Adding operational realism to developmental testing is very likely to help discover these flaws earlier in the development process. Another benefit of adding operational realism to developmental testing is that it provides a closer connection between developmental and operational testing, thereby facilitating combining information between the two forms of testing.

We also note that operational testing as currently done is typically too short to be able to discover many reliability deficiencies, such as fatigue. The time for developmental testing is also typically too short to find some of these flaws. These weaknesses in the current testing approach motivate the discussion below on accelerated testing, which, when properly implemented, can effectively expedite the discovery of design flaws.

A later report (National Research Council, 2006:15) noted that continuous testing is especially appropriate for systems that are acquired in stages, as one "learns about strengths and weaknesses of newly added capabilities or (sub) systems, and uses the results to improve overall system performance." This report also recommended that DOD documents and processes be revised "to explicitly recognize and accommodate [this] framework" (p. 3) so that the testing community is engaged in a joint effort to learn about and improve a system's performance. Although such formal changes have not been made, practices within DOD appear to be moving in this direction, one that is consistent with commercial industry practices.

There are a number of challenges in implementing the above paradigm. Test data from various sources need to be readily available, including field data from similar systems, data from previous stages of development, contractor data, developmental data, and data from modeling and simulation. Information from these sources can then be combined and exploited for effective test planning, design, analysis, and decision making. There are, however, major obstacles to meeting the challenges and accomplishing this approach in DOD: lack of data archives (see discussion below); use of multiple databases (with their own formats and incompatibilities); lack of access to data; and perhaps most importantly, lack of an incentive structure that emphasizes early detection of faults and sharing of information. As noted in the NRC report (2006:19): "incentives need to be put in place to support the process of learning and discovery of design inadequacies and failure modes early." In addition, the NRC recommended that DOD require that contractors share all relevant data on system performance and results of modeling and simulation developed under government contracts. Similarly, Adolph et al. (2008:219) noted: "Sharing and access to all appropriate system-level and selected component-level test and model data by government DT [developmental testing] and OT [operational testing] organizations" should be required in defense contracts. Despite these recommendations, there has been a lack of progress in this key area.

COMBINING INFORMATION

The importance of collecting and using all available data for effective decision making has been emphasized in several NRC reports.[1] Furthermore, it was the major focus of a subsequent report (National Research Council, 2004). Chapter 2 in that report deals with combining information to improve test planning and test design as well as analysis, and Chapter 3 discusses methods and examples related to reliability and suitability assessment. There is also an extensive statistical literature on this topic; in particular, an earlier NRC report (1992) is still a very useful reference.

Our contribution in this section is to provide some concrete ideas on how to parametrize the test space in order to improve test design and to combine results from different testing environments.[2]

A defense system is typically designed with some specific missions in mind. These missions can be characterized (at least partially) by variables that describe the environment of use (temperature, precipitation, wind speed, day/night, terrain, speed during use, weight of cargo, etc.). Other relevant factors include presence of countermeasures and enemy systems and the amount of training that the test personnel will have (which can vary widely from the so-called golden crews to the amount of training users will receive when a system is fielded). These factors may be ordered categorical variables or continuous variables. All possible combinations of these factors characterize the intended operational environment and hence the test space. These characterizations will often be incomplete in some respects since there may be some nominal (unordered) factors or some nuisance or noise factors that cannot be fully captured. The more effort that is placed in identifying and characterizing this space, the more efficient the testing program will be.

Both operational and developmental tests can be viewed as points in this space. Operational testing will use typical scenarios in the field and so may fall in the middle region in the test space (at least for some of the factors). Often, a systematic approach, such as statistical design of experi-

---

[1]See Recommendation 7.8 in National Research Council (1998:120) and Recommendation 2 in National Research Council (2003:53).

[2]The National Research Council (2006:18) report discussed such a test space: "We think that for test purposes, 'edge of the envelope' can be defined fairly rigorously. The space of conceivable military scenarios for operational testing includes a number of uncontrollable dimensions (e.g., environmental characteristics, potential missions, threat objectives and characteristics, etc.), and these dimensions can be usefully parameterized to identify the edge of the envelope." Bonder (1999) discusses parametric operational situation (POS) space formulation: "Each point in this space represents an operational situation that U.S. forces might have to be deployed to and operate in. Some of these situations are more stressful than others."

ments, is used to select the combinations of factor settings. Developmental testing is more ad hoc and will not examine the space systematically. Furthermore, it is likely to be based on more extreme scenarios, or what is often referred to as testing at the edge of the envelope.

Most of the operational test studies that we have seen are simple analyses that do not model the behavior of the factors over the test space. There is clearly some value in such analyses that do not make any assumptions and treat all the factors as nominal. But it would be very useful to also conduct additional analyses in which the effects of the factors are modeled parametrically (fitting parametric functions). Such analyses will allow a framework in which data from developmental tests (which may be isolated points in the test space) can be combined with data from operational tests to improve the information. Of course, part of the exploratory analysis will include checking for consistency among the developmental testing, operational testing, and other types of data, both empirically using extrapolations and using knowledge of the similarities and differences in the testing environments—and even for components and subsystems when available. If developmental testing includes scenarios at the edge of the envelope, the data can be interpolated to check for consistency with operational test data before they are combined. This framework also allows for the use of sequential testing during developmental testing with the aim of collecting more information in areas of the test space in which there are higher levels of uncertainty.

The panel recognizes that there are inherent dangers in combining data across heterogeneous sources without carefully considering the differences in the data sources and the reasons for the differences. Moreover, the ideas described here may not be applicable in all situations. For example, developmental test data may often be available only on components or subsystems. Nevertheless, it is important to examine different ideas on how to effectively combine data and effectively use test resources.

ACCELERATED TESTING

As the term suggests, accelerated testing involves conducting tests at conditions that are quite different from the operating conditions. Testing at the edge of the envelope, discussed above, can be viewed as one example. The discussion in this section deals mainly with reliability testing for suitability assessment.

The main goal in accelerated testing is to induce failures or degrade performance rapidly. Highly accelerated tests are commonly used by reliability engineers to identify failure modes. We focus here on the use of moderate acceleration regimens to estimate product or system reliability.

(An important caveat in these situations is that the acceleration should not induce failure modes that would not occur during normal operation.) Accelerated tests have been used extensively in industry. They are needed to estimate the reliability of highly reliable components or systems since few failures will occur during the (short) test phase of product development.

There are two common types of acceleration schemes: (1) increasing usage rate and reducing idle time and (2) using higher stress levels, such as temperature, voltage, humidity, and pressure. In the latter case, the appropriate stress factor(s) will depend on the component and failure mode of interest—corrosion, fatigue, mechanical wear, etc. There is an extensive discussion of stress factors corresponding to different types of failure mechanisms in the engineering literature.[3]

There is also considerable literature on the planning, design, and analysis of accelerated testing for life tests, where the outcome is lifetime data. The approach has also been used with degradation data (continuous measures of performance) although this literature is not as extensive (see Meeker and Escobar, 1998:Chs. 13, 21). Accelerated testing relies critically on the use of models to extrapolate the test results to normal use conditions. The literature emphasizes the need for using subject-matter knowledge and caution in extrapolating and suggests the use of extensive sensitivity analyses to assess the effects of using different models.

Accelerated testing is well known in the reliability community, and the panel expects that it is used extensively by defense contractors. However, given the inherent assumptions involved in these studies, it would be desirable for testers from DOD to either participate in their planning and analyses or have access to the test schemes in advance. Accelerated testing can and should play a prominent role in suitability assessment by DOD.

SOFTWARE SYSTEMS

Software systems are a major part of defense acquisition programs, either as exclusive systems or as critical parts of hardware systems. Software problems are also ubiquitous in poorly performing defense systems.[4] Although the use of processes such as agile development may lead to higher software quality, testing will remain crucially important. There is a substantial literature on software testing, and so we do not provide

---

[3]For an example, see *Reliability, Life Testing and the Prediction of Service Lives: For Engineers and Scientists* (Saunders, 2007).

[4]For example, see the report of the Defense Science Board, Task Force on Defense Software (2000).

an overview here. In particular, the NRC (2003:Ch. 3) has described techniques for software testing and related issues, including model-based testing, Markov-Chain usage models, and the use of combinatorial experimental designs.

There are some unique challenges with embedded systems, in which the software is embedded in hardware and has limited functionality (e.g., a GPS receiver) or is intended to react to a wide range of stimuli, such as the avionics for a jet fighter. These and other factors will determine if the software should be considered as simply a component of the full system during either developmental or operational testing or if the software needs to be tested separately from the remainder of the system. There is only a limited literature on the testing of embedded systems (but see Bringmann and Kramer, 2008, for some possibilities).

6

Communication, Resources,
and Infrastructure

Several aspects of infrastructure, expertise, and acquisition processes at the U.S. Department of Defense (DOD) hamper the application of best engineering practices. In this chapter we consider the importance of communication among the different teams involved in testing and development, data archiving, the use of feedback loops, and available systems engineering capabilities.

COMMUNICATION AND COLLABORATION AMONG REQUIREMENTS SETTING, DESIGN, AND TESTING

Conclusion 1 highlights the need for early and clear communications about requirements. In addition, industry representatives at the workshop stressed the importance of collaboration and communication among customers and program developers as well as participants across all aspects of system development and testing. Such collaboration is essential to avoiding long, costly, and unsuccessful product development programs. The drivers of unsuccessful commercial programs included the following features that panel members noted to be common to many troubled acquisition programs in DOD.[1]

---

[1]See Zyburt's presentation at http://www7.nationalacademies.org/cnstat/Presentations %20Main%20Page.html [November 2011].

- ever-changing program targets and functional objectives;
- providing inadequate or improper requirements to supplier or internal design group;
- lack of agreement on pass/fail criteria;
- late or no bench testing, which leads to a complete system becoming a "discovery property" rather than a "validation property"; and
- robust developed technologies are not "plugged in" to a program, and consequently advanced engineering and concept design occurs along the critical path of the program timeline versus offline.

Leading industrial companies have established programs to promote higher levels of collaboration among suppliers, manufacturers, customers, service organizations, and the ultimate users of the product. For example, Toyota Motor Company had two awards (excellence and superiority) to promote collaboration with and friendly competition among suppliers; see Box et al. (1988).

The recent Center for Naval Analysis (2009) study on the successful F/A-18E/F Super Hornet Development Program (mentioned above) reported on one of the few (recent) DOD development programs completed on time, within initial cost/required funding estimates, and meeting or exceeding all performance parameters—an outcome that resulted from close collaboration.[2] Close collaboration has also existed in earlier successful DOD programs, such as the nuclear attack submarine and ballistic missile submarine programs; it appears to be rare in recent DOD acquisition programs. Senior DOD acquisition executives in the Office of the Secretary and the military departments have the authority to require such close collaboration in the programs they oversee, but of late they have rarely required it or enforced it among the various groups critical to program success. The lack of coordinated efforts, particularly in the early stages of requirements and systems development, has contributed

---

[2]The program participants emphasized the importance of collaboration to achieve success. In this case, there was excellent partnership among the government program management office, the contractor program management office personnel and customer representatives from the Navy and the Office of the Secretary of Defense (OSD), the requirements community, the developmental test authority, the operational test agency, and a cost analysis improvement group. Several attributes were key in the process, which include the following:

- The program team took the time to get the requirements vetted and understood by all and revalidated those requirements every year.
- All team members were willing to work in an open and sharing manner—one team, one set of tools.
- The program team took the time to get the program planning right from the start—and executed the plan in an open and sharing way.
- There was disciplined change control throughout.

to the long-term detriment, and sometimes the cancellation, of a number of acquisition programs.

Collaboration can also be considered from the perspective of sharing information. Previous National Research Council (NRC) studies have emphasized the importance of using all available test information to improve operational evaluations, particularly with the use of evolutionary acquisition techniques. For example, National Research Council (2006:22) recommended that "the USD (AT&L) [Under Secretary of Defense for Acquisition, Technology, and Logistics] should develop and implement policies, procedures, and rules to . . . to share all relevant data on systems performance and the results of modeling and simulation . . . to assist in system evaluation and development." It is not possible to use and integrate information from the various sources without good collaboration and sharing of models and data across all of the important testing events.

DATA ARCHIVING

Conclusion 7: A data archive with information on developmental and operational test and field data will provide a common framework for discussions on requirements and priorities for development. In addition, it can be used to expedite the identification of and correction of design flaws. Given the expenses and complexity in developing such an archive, it is important that the benefits of a data archive be adequately demonstrated to support development.

Several previous NRC reports have also discussed this important topic,[3] but it has not received any noticeable attention in DOD. The collection and analysis of data on test and field performance, including warranty data, is a standard feature in commercial industries. In the DOD context, it is also important to retain information about test suites (by both contractors and DOD). In fact, it would be useful to require, through contractual obligation, that detailed information on tests carried out by contractors be provided to DOD. The panel does not make this suggestion lightly, as providing access to such test data is a large undertaking.

An archival test and field performance database could inform system developers as to the capabilities of components that had been used in fielded systems. Such a database could be extremely useful for requirements setting for future related systems. Furthermore, by capturing field

---

[3]Key among them are (1) Recommendation 3.3 in National Research Council (1998:49); Recommendation 2 in National Research Council (2003:3); National Research Council (2004:61); and Recommendation 6 in National Research Council (2006:37).

performance data, test scenarios can be selected to determine whether problems in a previously released system had been addressed in the most recent stage of development. Such a database could also help answer very broad questions about which development practices are most effective, cost impacts and trajectories, and what can be done to reduce development and acquisition costs. A key example is whether additional testing in development reduces life-cycle costs.

Such a database, if it included data on the performance of fielded systems, could support analyses similar to those of warranty systems in the commercial world. As Gilmore (2010) notes, DOD spends a substantial amount of its acquisition budget on operations and support. For example, for ground combat systems, the cost is 73 percent. A key driver of this cost is the poor reliability performance of the system and the resulting costs for replacement parts. A data archive could support analysis to control and manage a considerable fraction of operations and support costs by revealing and quickly fixing system deficiencies through a failure mode, effects, and criticality analysis, and a failure reporting, analysis, and corrective action system supported by such data collection.

This database would need to be easily accessible by program managers and testers. It is important for everyone to work from the same database so that requirements, specifications, and later assertions of reliability and effectiveness based on archived test results and results from modeling and simulation can be compared and contrasted. The speakers at the workshop insisted that developers in industry know the historical performance of components or subsystems that are included in the system in question, and so they can then anticipate problems in development and work to prevent them. Therefore, it is extremely important to also include contractor test results in such an archive, since that is the only way the full history of performance can be represented.

> **Recommendation 3: The U.S. Department of Defense (DOD) should create a defense system data archive, containing developmental test, operational test, and field performance data from both contractors and the government. Such an archive would achieve several important objectives in the development of defense systems:**
>
> - **substantially increase DOD's ability to produce more feasible requirements,**
> - **support early detection of system defects,**
> - **improve developmental and operational test design, and**
> - **improve modeling and simulation through better model validation.**

Given these important benefits, DOD should initiate plans to begin creation of a defense system data archive. Some issues that need immediate resolution include (a) whether the archive should be DOD-wide or should be stratified by type of system to limit its size, (b) what data are to be included and how the data elements should be represented to facilitate linkages of related systems, and (c) what data-based management structure is used. In designing this archive, a flexible architecture should be used so that if the archive is initially limited to a subset of the data sources listed here due to budgetary considerations, the archive can be readily expanded over time to include the remaining sources.

Specification of how such a database should be constructed and what it should contain are beyond the scope of this study. DOD currently has multiple databases that have been developed in the different services for different types of systems to satisfy various needs. They represent some aspects of the database we are describing. There are databases with developmental test data, databases that collect operational test data, databases with modeling and simulation results, and databases with field performance data. Unfortunately, in most cases, these databases are not compatible with each other. Perhaps an initial approach to the development of a test and field data archive would be to institute linkages that allow the combination of system-specific information across the existing databases.

A key reason for the lack of progress in this area is the incentive structure in the DOD acquisition environment. Individual programs do not obtain any immediate benefit from committing resources for the development and maintenance of data archives beyond their own program for the common good. So the first step would be for DOD to recognize the advantages of building and maintaining such a database and exploring how a data archive would be funded. With this recognition, the panel suggests that DOD could commission a committee of people with expertise in database management and people with experience in program development to propose concrete recommendations.

FEEDBACK LOOPS

Conclusion 8: Feedback loops can significantly improve system development by improving developmental and operational test design and improving the use of modeling and simulation. Feedback systems can function similarly to warranty management systems that have proved essential to the automotive industry. To develop feasible requirements, understanding how components installed in related systems have performed when fielded is extremely useful in understanding their limitations for possibly more stressful use in a proposed system. To support such feedback loops, data on field

performance, test data, and results from modeling and simulation must be easily accessible, which highlights the necessity for a test and field data archive.

Field performance data are the ultimate indicators of how well a system is functioning in actual operational conditions. By field performance data, we include all the circumstances that can have an impact on the quality of the components, subsystems, and a system itself. These circumstances include all relevant pre- and postdeployment activities, including transportation, maintenance, implementation, and storage. They could also include training data, if such data are collected objectively.

Such data should be used to better understand the strengths and weaknesses of newly fielded systems and can be used in feedback loops. They can indicate when components or subsystems should be modified because of inferior effectiveness or suitability, and they can be used to identify for which missions the current system will work. For instance, if a system exhibits poor reliability in certain stressful scenarios of use, say, while carrying loads of more than a given weight, and if the reliability of the system under such conditions cannot be easily or quickly improved, the information can support a decision not to use the system for such missions (if alternatives are available). And, if the reliability of the relevant component can be improved with a redesign, the information can be used to support arguments for such a redesign.

Design flaws that are identified in fielded systems can also be evidence of failure in the testing process. For instance, inferior reliability of a system under heavy loads is likely to be an indication that those weights were not used during developmental and operational testing. The reason for such an omission can then be examined, and the process for selection of experimental designs can be improved. Also, field performance data can provide information on the validity of any modeling and simulation that were used to assess operational performance. For example, if modeling and simulation were used to extrapolate from light loads instead of actual physical testing, the validity of the use of modeling and simulation can be examined and the process for validating modeling and simulation can then be improved.

The National Research Council (2003) noted two significant benefits of feedback loops: field performance data can be used to help estimate total life-cycle costs of a newly fielded system, and, in spiral development, effective feedback processes can identify enhancements that will improve the effectiveness and suitability of later stages of development. Improving the quality and timeliness of this feedback is important in being able to respond to rapid changes in threat environments.

It is the panel's understanding that such a feedback loop currently operates only when a system is underperforming in a dramatic way. Instead, such analysis and feedback should be routine. Although DOD does collect field performance data, they are of highly varying quality and are not archived in a way that facilitates analysis (see discussion in Chapter 5). To varying degrees, the services do use a deficiency reporting process as a feedback mechanism during developmental programs, starting with the design review and continuing through testing. Deficiencies are categorized to identify the relative importance and urgency of a response. For example, in the Air Force, the stated purpose of the deficiency reporting investigation and resolution process is to provide "a means of identifying deficiencies, resolving those deficiencies within the bounds of program resources and the appropriate acceptance of risk for those deficiencies that cannot be resolved in a timely manner" (U.S. Air Force, 2007:1-1). However, the process has mostly been allowed to atrophy in the past 15 years, for several reasons, most notably the services' lack of participation in developmental testing and the need to ignore all but the most critical deficiencies that are identified because of a lack of funds to take corrective actions.

> **Recommendation 4: After a test and field data archive has been established, the Under Secretary of Defense for Acquisition, Technology, and Logistics and the acquisition executives in the military services should lead a U.S. Department of Defense (DOD) effort to develop feedback loops on improving fielded systems and on better understanding tactics of use of fielded systems. The DOD acquisition and testing communities should also learn to use feedback loops to improve the process of system development, to improve developmental and operational test schemes, and to improve any modeling and simulation used to assess operational performance.**

SYSTEMS ENGINEERING CAPABILITIES IN DOD

Conclusion 9: The U.S. Department of Defense has lost much of its expertise in all aspects of systems engineering in recent years. It is important to have in-house capability in the critical areas relating to the design, development, and operation of major types of systems and subsystems. One such area is expertise in model-based design tools.

Some of the speakers at the workshop noted that commercial companies stress the importance of systems engineering expertise. This expertise is key not only for system development but also for requirements setting, model development, and testing. In contrast, Adolph (2010:51) notes that in DOD: "The manpower reductions mandated by Congress in the late 1990s, followed by excessive additional services-directed reductions, have decimated the program office engineering and test support workforce as well as DOD government test organization personnel." In addition, Adolph et al. (2008:220), summarizing a 2008 report of the Defense Science Board, state: "A second and related priority is to ensure that government organizations reconstitute a cadre of experienced test and evaluation, engineering, and RAM personnel to support the acquisition process."

In order to improve its test and development process, DOD will have to reverse this trend. It appears that DOD has recognized this problem and is taking steps to rectify it. The panel applauds this effort, but we emphasize that, even with a dedicated and sustained effort, it will take a take a decade or more to have the capabilities that DOD had in the early 1990s. Therefore, DOD should examine short-term use of contractors, academics, employees of national laboratories, etc. so that many of the recommendations in this and other studies can be implemented in a timely manner. The problem of systems engineering capability is also complicated by the reduced numbers of U.S. citizens who are acquiring engineering degrees. DOD should examine creative alternatives, including ways to engage noncitizen engineers on DOD acquisition programs, temporary employment opportunities, fellowships, internships, and sabbaticals of various kinds.

7

Organizational Structures and Related Issues

In this chapter, we focus on two topics: the lack of enforcement of existing U.S. Department of Defense (DOD) guidelines and procedures and the role of the program manager in the acquisition process.

ENFORCEMENT OF DOD DIRECTIVES AND PROCEDURES

Conclusion 10: Many of the critical problems in the U.S. Department of Defense acquisition can be attributed to the lack of enforcement of existing directives and procedures rather than to deficiencies in them or the need for new ones.

Christie (2011) discussed this issue and pointed to several aspects of it:

1. a lack of discipline in decision making concerning advancement of programs through the defense acquisitions milestone system;
2. unfortunate incentives that result in overly optimistic initial statements of systems requirements as well as optimism regarding the expeditiousness of development and the costs of production and fielding;
3. failure to rigorously demonstrate, through empirical testing, the required technological maturity of a component or subsystem before each major milestone decision point;

4. failure to first establish and then to carry along event-based strategies—instead employing schedule-based strategies—and failure to use strict pass/fail criteria for each phase of development;
5. failure to carry out continuous, independent assessments of the effectiveness and suitability of defense systems in development from initial development through the various stages of testing and production, extending to early introduction to the field; and
6. failure to use feedback loops to inform the broad acquisition community as to when acquisition methods have worked and when they have failed so that all can learn from others' experiences.

We discuss several of these factors throughout this report.

The following actions, some of which are discussed in the report, can help ameliorate these problems:

- Competitive prototype development and testing should be a strict prerequisite for any new development program prior to entry into engineering and manufacturing development.
- Emphasis should be on an event-based strategy, rather than a schedule-based strategy, with meaningful and realistic pass/fail criteria for each stage of development. In particular, systems should not be allowed to proceed to operational testing unless that step is supported by developmental test performance that strongly anticipates that the system will pass; such a determination can be greatly aided through the conduct of a rigorous operational test readiness review.
- Use of continuous and independent evaluation tracking of each program through the stages of development, testing, and production should be required. These assessments should rely heavily on empirical tests and should focus on those capabilities that were the basis for program approval.

Problems with suitability performance of defense systems are just as widespread, and the Defense Science Board (2008) made the following recommendations for remedying them:

1. Identify reliability, availability, and maintainability (RAM) requirements during the joint capabilities integration development system process, and "incorporate them in the request for proposal as a mandatory contractual requirement" (p. 6).
2. When evaluating proposals, evaluate the bidder's approaches to satisfying RAM requirements.

3. Include a robust reliability growth program as a mandatory contractual requirement and document progress during each major program review.
4. Include a credible reliability assessment as part of the various stages of technical review.
5. Raise the importance of achieving RAM-related goals in the responsibilities of program managers.
6. Develop a standard for RAM development and testing.
7. Increase the available expertise in reliability engineering.

THE ROLE OF A PROGRAM MANAGER

The concept of having a strong project manager, sometimes called a chief engineer, was pioneered by Honda. It was pervasive in Japan as early as the 1980s (Box et al., 1988) and has become a standard practice in the automotive industry in the United States. The program manager is appointed early in an acquisition process, as soon as product feasibility is demonstrated through a successful market study. The program manager's responsibility covers the entire spectrum, from planning, design, development, and manufacturing to even initial phases of sales and field support.

The organizational structure of the teams and implementation details vary across companies, but there is usually continuity with a few team members going from one phase to be part of the team for the next phase. This practice ensures a smooth transition as well as transfer of knowledge. But the key person is the program manager, who is fully responsible and accountable for all phases of the product realization process. If the system has difficulties in development, such as delays or cost increases, or if the system underperforms when fielded, final responsibility lies with the program manager. A strong program manager has the authority to assemble the right team members from inside the corporation; to hire or contract other needed skills; to approve final designs, requirements, vendors and suppliers; and to set the final schedule. Input from all divisions—including sales, marketing, dealers, and manufacturing plants is actively solicited—but the final decisions are made by the program manager. Other industries, besides automotive, also use a similar concept of having a single person in charge of the entire product realization process.

The same activities occur in DOD programs in the broader context of the acquisition cycle. Every program is managed sequentially through phases, all followed by major milestones in which decision makers approve or disapprove of the acquisition strategy before the program moves to the next phase of development.[1]

For DOD programs, however, there are two people with the title of program manager. One is appointed by the defense contractor and generally remains in charge for an extended period of time. The other is designated by DOD: that person is typically a military officer whose chief responsibility is to manage the system development to the next milestone, but his or her tenure is often shorter than the time span between milestones. Tenures have been lengthening of late, but they are still much shorter than development times. The DOD norm is that after a program manager's tour is concluded, the person is generally promoted and replaced, and the status of the acquisition program during that person's tenure is not carefully assessed (as it often is in industry). The short tenure and lack of accountability lead to disincentives. For example, there is no motivation for a program manager to be comprehensive in discovering system defects or design flaws in the early stages of system development. Furthermore, given the turnover, any deficiencies are unlikely to be attributed to the efforts of a specific program manager. This approach is in stark contrast with industry, which has more stability and the right incentive structure, which includes being aggressive about finding system defects as early in system development as possible.[2]

The panel recognizes the challenges associated with program management and does not expect any significant changes to the present system of short-term rotations of military officers as program managers. Nevertheless, we believe that DOD should explore ways to provide more stability, and thereby accountability, to the project management process. Two possibilities include (1) developing a new civilian position in which a person can serve as deputy to each of the program managers and whose tenure spans a substantial portion of the system development cycle, and (2) appointing a deputy program manager at each milestone with the expectation that the deputy will be promoted to program manager.[3] Of course, the problem with the incentive structure for program managers will remain, and it is unclear how they would respond to a civilian or to a deputy.

Regardless of these possibilities, the panel believes that there has to be an independent third-party assessment of ACAT (acquisition category)

---

[1]See Appendix B for an overview of the defense acquisition process; see U.S. General Accounting Office (1998) for the role of a program manager.

[2]For an analogous discussion of space systems, see Defense Science Board (2003).

[3]Bell (2008:277) argues: "On the other hand, PMs and their PMOs have to start taking the long-term or enterprise view. That is, it is **not** OK to delay the discovery of technical, schedule, or budget problems until a future PM has no choice but to acknowledge them. PMs need to be rewarded for solving problems, not for postponing them."

I systems whenever a program manager leaves. This assessment needs to be carried out by personnel who are outside the influence of the services and, in particular, external to the acquisition contract for the program. This assessment would allow for the progress of the system under that program manager to be determined objectively. Moreover, the success of each new program manager should be assessed only on the basis of the status and progress from the point of transition. Such an assessment may also change the incentive structure: each program manager will have an incentive to discover design flaws because the improvement of the system under his or her tenure would now be directly evaluated.

We do not offer any suggestions on how the performance of program managers should be assessed if they failed to discover design flaws and system defects. Also, guidelines would have to be developed on how problems from earlier stages of development—for example, that a system's performance was not comprehensively tested or discovered flaws were left unaddressed—would affect the assessment of subsequent program managers.

> **Recommendation 5: The Under Secretary of Defense for Acquisition, Technology, and Logistics should provide for an independent evaluation of the progress of acquisition category I (ACAT I) systems in development when there is a change in program manager. This evaluation should include a review by the Office of the Secretary of Defense (OSD), complemented by independent scientific expertise as needed, to address outstanding technical manufacturing and capability issues, to assess the progress of a defense system under the previous program manager, and to ensure that the new program manager is fully informed of and calibrated to present and likely future OSD concerns.**

Clearly, there are many details and challenges associated with developing and implementing this recommendation. These are beyond the panel's scope and expertise, but we conclude there are systemic problems with the current system of program management that are obstacles to the implementation of efficient practices.

References

Adolph, P. (2010). The acquisition/test process: What went wrong and how to fix the process. *ITEA Journal* 31:49-56.

Adolph, P., DiPetto, C.S., and Seglie, E.T. (2008). Defense Science Board task force developmental test and evaluation study results. *ITEA Journal* 29:215-221.

Bell, W.D. (2008). The 800-pound (364 kg) gorilla. *ITEA Journal* 29:275-277.

Bonder, S. (2000). *Versatility Planning: An Idea Whose Time Has Come—Again!* Steinhardt lecture presented at the Institute for Operations Research and the Management Sciences Conference, May 9, Salt Lake City, UT.

Box, G., Kackar, R., Nair, V., Phadke, M., Shoemaker, A., and Wu, C.F.J. (1988). Quality practices in Japan. *Quality Progress* March:37-41.

Bringmann, E., and Kramer, A. (2008). *Model-Based Testing of Automotive Systems.* Presented at the Software Testing, Verification, and Validation, 2008 1st International Conference, April 9-11, Lillehammer, Norway. Abstract available: http://ieeexplore.ieee.org/xpls/abs_all.jsp?arnumber=4539577&tag=1 [December 2011].

Center for Naval Analysis. (2009). *The F/A-18E/F Super Hornet Development Program: Could This Success Story Be Repeated Today?* G. Christle and D. Davis. CNA Analysis & Solutions report CRM D0021571.A2/Final. Alexandria, VA: Center for Naval Analysis.

Christie, T. (2011). Essay 10: Developing, buying, and fielding superior weapon systems. In W.T. Wheeler (Ed.), *The Pentagon Labyrinth: Ten Short Essays to Help You Through It.* Washington, DC: Center for Defense Information, World Security Institute.

Defense Science Board. (2000). *Report of the Defense Science Board Task Force on Defense Software.* Washington, DC: Office of the Under Secretary of Defense, U.S. Department of Defense.

Defense Science Board. (2003). *Acquisition of National Security Space Programs.* Washington, DC: Defense Science Board/Air Force Scientific Advisory Board Joint Task Force; Office of the Under Secretary of Defense for Acquisition, Technology, and Logistics.

Defense Science Board. (2008). *Report of the Defense Science Board Task Force on Developmental Test and Evaluation.* Office of the Secretary of Defense for Acquisition, Technology, and Logistics. Available: http://www.acq.osd.mil/dsb/reports/ADA482504.pdf [July 2011].

Gilmore, J.M. (2010). *Director, Operational Test and Evaluation FY 2010 Annual Report.* Available: http://www.defensenews.com/static/defense_fy2010_dote_annual_report.pdf [December 2011].

Institute for Defense Analyses. (1991). *The Role of the Office of the Secretary of Defense in the Defense Acquisition Process.* Prepared for the Office of the Under Secretary of Defense for Acquisition. Alexandria, VA: Institute for Defense Analyses. Available: http://pogoarchives.org/labyrinth/10/01.pdf [December 2011].

Mathworks®. (2009). *General Motors Develops Two-Mode Hybrid Powertrain with Model-Based Design.* Press release. Available: http://www.mathworks.com/company/pressroom/General-Motors-Develops-Two-Mode-Hybrid-Powertrain-With-Model-Based-Design.html [July 2011].

Meeker, W.Q., and Escobar, L.A. (1998), *Statistical Methods for Reliability Data.* New York: Wiley.

National Defense Industrial Association. (2011). *Final Report of Model Based Engineering (MBE) Subcommittee.* Arlington, VA: NDIA Systems Engineering Division, M&S Committee.

National Research Council. (1992). *Combining Information: Statistical Issues and Opportunities for Research.* Panel on Statistical Issues and Opportunities for Research in the Combination of Information. Committee on Applied and Theoretical Statistics. Board on Mathematical Sciences. Commission on Physical Sciences, Mathematics, and Application. Washington, DC: National Academy Press.

National Research Council. (1998). *Statistics, Testing, and Defense Acquisition: New Approaches and Methodological Improvement.* Panel on Statistical Methods for Testing and Evaluation Defense Systems, M. Cohen, J. Rolph, and D. Steffey, Eds. Committee on National Statistics, Commission on Behavioral and Social Sciences and Education. Washington, DC: National Academy Press.

National Research Council. (2003). *Innovations in Software Engineering for Defense Systems*; Oversight Committee for the Workshop on Statistical Methods in Software Engineering for Defense Systems. S.R. Dalal, J.H. Poore, and M.L. Cohen, Eds. Committee on National Statistics, Division of Behavioral and Social Sciences and Education. Washington, DC: The National Academies Press.

National Research Council. (2004). *Improved Operational Testing and Evaluation and Methods of Combining Information for the Stryker Family of Vehicles and Related Army Systems. Phase II Report.* Panel on Operational Test Design and Evaluation of the Interim Armored Vehicle, Committee on National Statistics, Division of Behavioral and Social Sciences and Education. Washington, DC: The National Academies Press.

National Research Council. (2006). *Testing of Defense Systems in an Evolutionary Acquisition Environment.* Oversight Committee for the Workshop on Testing for Dynamic Acquisition of Defense Systems. V. Nair and M.L. Cohen, Eds. Committee on National Statistics, Division of Behavioral and Social Sciences and Education. Washington, DC: The National Academies Press.

National Research Council. (2008). *Pre-Milestone A and Early-Phase Systems Engineering: A Retrospective Review and Benefits for Future Air Force Acquisition*, Air Force Studies Board, Division on Engineering and Physical Sciences. Washington, DC: The National Academies Press.

National Research Council. (2010a). *Achieving Effective Acquisition of Information Technology in the Department of Defense*. Committee on Improving Processes and Policies for the Acquisition and Test of Information Technologies in the Department of Defense. Computer Science and Telecommunications Board, Division on Engineering and Physical Sciences. Washington, DC: The National Academies Press.

National Research Council. (2010b). *Evaluation of U.S. Air Force Preacquisition Technology Development*. Committee on Evaluation of U.S. Air Force Preacquisition Technology Development. Air Force Studies Board, Division on Engineering and Physical Sciences. Washington, DC: The National Academies Press.

Nelson, W. (1990). *Accelerated Testing Statistical Models, Test Plans and Data Analysis*. New York: Wiley.

Saunders, S.C. (2007). *Reliability, Life Testing and the Prediction of Service Lives: For Engineers and Scientists*. New York: Springer.

Schwartz, M. (2010). *Defense Acquisitions: How DoD Acquires Weapon Systems and Recent Efforts to Reform the Process*. Congressional Research Service, 7-5700, RL34026. Available: http://www.fas.org/sgp/crs/natsec/RL34026.pdf [December 2011].

Thompson, F. (1992). Deregulating defense acquisition. *Political Science Quarterly* 107(4): 727-749.

U.S. Air Force. (2007). *Technical Manual: USAF Deficiency Reporting, Investigation, and Resolution*. T.O. 00-35D-54. Available: http://www.tinker.af.mil/shared/media/document/AFD-061214-036.pdf [December 2011].

U.S. Department of Defense. (2008). *Operation of the Defense Acquisition System*. U.S. DOD Instruction (DODI) 5000.02. Available: http://www.dtic.mil/whs/directives/corres/pdf/500002p.pdf [July 2011].

U.S. Department of Defense. (2010). *Technology Maturity and Integration Risk of Critical Technologies for CY 2009*. Washington, DC: U.S. Department of Defense.

U.S. General Accounting Office. (1992). *Weapons Acquisition: A Rare Opportunity for Lasting Change*. GAO/ NSAID-93-15/December 1992. Available: http://www.gao.gov/products/NSIAD-93-15 [December 2011].

U.S. General Accounting Office. (1998). B*est Practices: Successful Application to Weapons Acquisitions Requires Changes in DoD's Environment*. GAO/NSAID-98-56. Available: http://www.gao.gov/products/NSIAD-98-56 [December 2011].

U.S. General Accounting Office. (1999). *Best Practices: Better Management of Technology Development Can Improve Weapon System Outcomes.* Report to the Chairman and Ranking Minority Member, Subcommittee on Readiness and Management Support, Committee on Armed Services, U.S. Senate. GAO/NSAID-99-162. Washington, DC: U.S. General Accounting Office.

U.S. Government Accountability Office. (2004). *Defense Acquisitions: The Army's Future Combat System's Features, Risks, and Alternatives*. GAO-04-635T. Available: http://www.gao.gov/new.items/d04635t.pdf [December 2011].

U.S. Government Accountability Office. (2006). *Best Practices: Stronger Practices Needed to Improve DOD Technology Transition Processes*. Report to Congressional Committees. GAO-06-883. Washington, DC: U.S. Government Accountability Office.

Appendix A

Workshop Agenda

Day 1: Friday, January 15, 2010

8:30 am   Introduction and Kickoff
               Vijay Nair, University of Michigan (*Committee Chair*)

**SESSION A: VIEWS FROM NONDEFENSE INDUSTRIES—PART I:
               SOFTWARE**
               Moderator: Alyson Wilson, Science and Technology Policy
               Institute, Institute for Defense Analyses (*Panel Member*)

               Speakers:
8:50 am   Donald Bollinger, Software Engineering, Hewlett-Packard
9:30 am   Sham Vaidya, Emerging Architecture and Technology, IBM

               Discussion:
10:10 am   Steve J. Hutchison, test and evaluation executive,
               Defense Information Systems Agency
10:20 am   Elaine Weyuker, AT&T Laboratories Research (*Panel
               Member*)
10:30 am   Open Discussion

10:40 am   Break

SESSION B: VIEWS FROM NONDEFENSE INDUSTRIES— PART II: HARDWARE
Moderator: John Christie, LMI (*Panel Member*)

Speakers:
11:00 am   Jeff Zyburt, independent consultant (formerly director of Testing, Chrysler)
11:40 am   Salim Momin, independent consultant (formerly with Motorola/Freescale)

12:30 pm   Lunch

Discussion:
1:30 pm   Dmitry Tananko, General Dynamics
1:40 pm   A. Blanton Godfrey, North Carolina State University (*Panel Member*)
1:50 pm   Floor Discussion

SESSION C: EXPERIENCES IN DOD AND DEFENSE INDUSTRIES
Moderator: Michael Cohen, CNSTAT (*Panel Staff*)

Speakers:
2:00 pm   Robin Pope, SAIC—The Future Combat System as a Case Study
2:50 pm   Michael Cushing, ATEC

Discussion:
3:30 pm   Pete Adolph, retired (*Panel Member*)
3:40 pm   Raj Kawlra, Chrysler (*Panel Member*)
3:50 pm   Floor Discussion

4:00 pm   Coffee

SESSION D: PANEL DISCUSSION
Moderator: John Rolph, University of Southern California (*Panel Member*)

4:15 pm   **Panelists:**
William McCarthy, DOT&E (formerly OPTEVFOR)
Steve Welby, director, Systems Engineering
Chris DiPetto, acting director, Development Test
Tom Christie, retired (*Panel Member*)
Peter Cherry, SAIC (*Panel Member*)

5:05 pm	Floor Discussion
5:30 pm	Adjourn

Day 2: Saturday, January 16, 2010

SESSION E: BRAINSTORMING SESSION

9:00 am	How to Find Failures Early in Development
9:45 am	How to Assess Technological Maturity
10:30 am	Break
10:45 am	How to Combine Information from Disparate Sources
11:30 am	Challenges Facing DOD in Implementing Industrial Best Practices and Other Issues
12:00 pm	Working Lunch
2:00 pm	Adjourn

Appendix B

Overview of the Defense Milestone System

This appendix presents a brief overview of the defense acquisition process.[1] The diagram below from the U.S. Department of Defense (2008:12 enclosure 2) depicts the development of defense systems as they proceed through the defense acquisition milestone system:

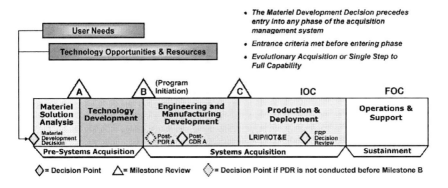

For the purposes of this report, it is generally sufficient to know the following rough outline of the process. Defense systems have to be justified as satisfying a specific military need. This occurs during the

---

[1]Some of this discussion relies on Schwartz (2010). For more details, see U.S. Department of Defense Instruction (DODI) 5000.02, *Operation of the Defense Acquisition System*. Available: http://www.dtic.mil/whs/directives/corres/pdf/500002p.pdf [November 2011].

first stage—materiel solution analysis. After that has been established, specific requirements are produced, a request for proposals is generated, a contract is awarded, work on a specific system begins, and the system passes Milestone A.

At this stage, each acquisition program is usually managed by an acquisition program office with an assigned program manager (PM). The PM is usually supported by a staff that can include engineers, logisticians, contracting officers, and various system specialists.

Milestone A is followed by technology development in which the necessary technologies for the system are identified and, if not sufficiently mature, are further developed. This phase is also when the reliability, availability, and maintainability strategy is developed. The technology development stage culminates in Milestone B.

This stage of development is complete, passing Milestone B, only when an affordable system is identified that is sufficiently mature in the relevant environment, and it has also been shown that the needed manufacturing processes are ready to produce prototypes at a reasonable cost. Passing Milestone B, a system enters into engineering and manufacturing development. In this stage, the components and subsystems are fully integrated into a complete system, and manufacturing processes are finalized. This phase is when a majority of the developmental testing takes place. Systems pass Milestone C when they have passed developmental testing and operational assessment, when they have shown interoperability and operational supportability, and when they have been shown to be affordably manufactured.

The next stage is production and deployment, which works with a small number of system prototypes, produced as part of low-rate initial production, and includes the most operationally relevant form of testing—operational testing. Systems that pass operational testing enter into full-rate production. The final stage of acquisition is operations and support, which oversees the continued use and improvement of the system through its lifetime of use.

Appendix C

Biographical Sketches of Panel Members and Staff

VIJAY NAIR (*Chair*) is Donald A. Darling professor of statistics and professor of industrial and operations engineering at the University of Michigan. His past experience includes 15 years as a research scientist at Bell Laboratories. He has a broad range of interests in statistical methodology and applications, especially in engineering statistics. He is president-elect of the International Statistical Institute and president of the International Society for Business and Industrial Statistics. He is a senior fellow of the Michigan Society of Fellows and a fellow of the American Association for the Advancement of Science, the American Society for Quality, the American Statistical Association, and the Institute of Mathematical Statistics. He is a former editor of *Technometrics* and the *International Statistical Review*. He has also served as chair of the board of trustees of the National Institute of Statistical Sciences. He has a Ph.D. in statistics from the University of California, Berkeley.

CHARLES E. (PETE) ADOLPH is an independent consultant with several decades of experience in testing and evaluation and acquisition management. He began his career with General Dynamics Convair as a flight test engineer at Edwards Air Force Base; served as a U.S. Air Force officer; held a variety of civilian engineering and systems acquisition, technical, and management positions with the Air Force; and served as the technical director, the senior civilian position at the Air Force Flight Test Center. He also held several positions in the Office of the Secretary of Defense, including director of test and evaluation in the Office of the Undersecretary of Defense for Acquisition and Technology. He also held senior management positions with Science Applications International Corporation. He received a B.S. in aeronautical engineering from St. Louis University, an M.S. in aeronautical and astronautical engineering from the University of Michigan, and an M.S. in systems management from the University of Southern California.

W. PETER CHERRY recently retired from his position as chief analyst at Science Applications International Corporation, where his research interests included the design, development, and test and evaluation of large-scale systems with emphasis on network centricity. He has focused on the development and application of operations research in the national security domain, primarily in the field of land combat. He contributed to the development and fielding of most of the major systems currently employed by the Army, ranging from the Patriot Missile System to the Apache helicopter, as well as the command, control, and intelligence systems currently in use. In addition, he contributed to the creation of the Army's Manpower Personnel and Human Factors and Training Program and to the Army's Embedded Training Initiative. He is a member of the National Academy of Engineering. He has a B.A. from the University of New Brunswick and an M.A. from the University of Toronto, both in mathematics. He also holds an M.S. and Ph.D. in industrial and operations engineering from the University of Michigan.

JOHN D. CHRISTIE is senior fellow at the Logistics Management Institute. He has an extensive background in U.S. Department of Defense acquisition policy, program analysis, and resource allocation, having served as director of acquisition policy and program integration for the Office of the Under Secretary of Defense for Acquisition. In that position, he prepared a comprehensive revision of all defense acquisition policies and procedures, resulting in the cancellation and consolidation of 500 prior separate issuances. He also prepared comprehensive acquisition program alternatives for the secretary of defense that resulted in budget reductions of billions of dollars. He has S.B., S.M., E.M.E., and Sc.D. degrees from the Massachusetts Institute of Technology, all in mechanical engineering.

THOMAS P. CHRISTIE, an independent consultant, last served as the director of operational test and evaluation for the U.S. Department of Defense (DOD). In that position, he advised and consulted with the secretary of defense and senior assistants in setting DOD policy and procedures for the testing of new weapon systems, weapons support systems, equipment, and munitions. In his career at DOD, he also served as director of program integration for the Office of the Under Secretary for Acquisitions and Technology and in the Office of the Assistant Secretary of Defense and in the Office of the Deputy Assistant Secretary of Defense for General Purpose Programs. Previously, he served as director of the weapon system analysis division at the Air Force Armament Laboratory at Eglin Air Force Base. In addition to his career at DOD, he worked on DOD weapons testing at the Institute for Defense Analyses. He holds a B.S. in mathematics from Spring Hill College and an M.S. in applied mathematics from New York University.

MICHAEL L. COHEN (*Study Director*) is a senior program officer for the Committee on National Statistics where he directs studies involving statistical methodology, in particular, on defense system testing and decennial census methodology. He has also recently worked on the prevention and treatment of missing data in clinical trials and data mining applied to counterterrorism. Formerly, he was a mathematical statistician at the Energy Information Administration, an assistant professor in the School of Public Affairs at the University of Maryland, and a visiting lecturer in the Department of Statistics at Princeton University. He is a fellow of the American Statistical Association. He has a B.S. in mathematics from the University of Michigan and an M.S. and a Ph.D. in statistics from Stanford University.

A. BLANTON GODFREY is dean and Joseph D. Moore professor of textile and apparel technology and management at the College of Textiles, North Carolina State University. His research interests include business management and new product development, quality and productivity management, strategic planning and deployment, experimental design, reliability, data analysis, and applied statistics. Previously, he was chair and chief executive officer of the Juran Institute, Inc., a management consulting, research, and training organization focused on quality management. He previously also served as head of the Quality Theory and Technology Department of AT&T Bell Laboratories. He is a fellow of the American Statistical Association and of the American Society for Quality and an elected member of Sigma Xi. He received a B.S. in physics from Virginia Polytechnic Institute and State University and an M.S. and a Ph.D. in statistics from Florida State University.

RAJ KAWLRA is director of dimensional management and strategies for manufacturing engineering at the Chrysler Group, LLC. His previous positions at Chrysler included director of manufacturing quality, with responsibility for quality systems, procedures, and processes, including vehicle assembly, powertrain, and stamping facilities. His prior work at

General Motors (GM) included serving as an adviser to the company's lean manufacturing core planning team and as the engineering group manager for the math-based quality systems at the GM Tech Center. His primary focus has been on the development of new technologies to improve quality and throughput. He received a B.S. in mechanical engineering from Banaras Hindu University in India; an M.S. in mechanical engineering from the University of Wisconsin, Madison; an M.S. in industrial engineering from the University of Illinois, Urbana-Champaign; and a Ph.D. in industrial and operations engineering from the University of Michigan, Ann Arbor.

JOHN E. ROLPH is professor of statistics at the Marshall School of Business of the University of Southern California, where he also holds appointments in the mathematics department and the law school. Previously, he was a statistician at the RAND Corporation and head of the statistical research and consulting group. His areas of expertise include statistics and public policy and empirical Bayes estimation. He is an elected member of the International Statistical Institute, a fellow of the American Statistical Association, a fellow of the Institute of Mathematical Statistics, and a lifetime national associate of the National Academies. He is a past editor of *CHANCE* magazine and has served in many other editorial capacities. He has a Ph.D. in statistics from the University of California, Berkeley.

ELAINE WEYUKER is a principal technical staff member at AT&T Laboratories. Previously, she was a professor of computer science at the Courant Institute of Mathematical Sciences of New York University. Her research interests are in software engineering, particularly software testing and reliability and software metrics. In each of the past 6 years, the *Journal of Systems and Software* has rated her as one of the top five software engineering researchers in the world. She is a member of the National Academy of Engineering and a fellow of the Institute of Electrical and Electronics Engineers and of the Association of Computing Machinery. She received an M.S.E. from the Moore School of Electrical Engineering at the University of Pennsylvania and a Ph.D. in computer science from Rutgers University.

MARION L. WILLIAMS is an adjunct research staff member at the Institute for Defense Analyses. Prior to this position, he served as chief scientist and technical director of the Air Force Operational Test and Evaluation Center. He previously served as an aerodynamicist at Sandia National Laboratories and as an adjunct professor in the electrical engineering department at the University of New Mexico. He has been a member of numerous scientific panels, including the Defense Science Board study on developmental test and evaluation and the Air Force Scientific Advisory Board studies on test and evaluation, modeling and simulation, and electronic warfare. Among his many awards are the Vance Wanner Award from the Military Operations Research Society, the Allan Matthews Award from the International Test and Evaluation Association, and the Air Force Association citation of honor. He received a B.S. in aeronautical engineering from Texas A&M University, an M.S. in mechanical engineering from the University of New Mexico, and a Ph.D. in industrial engineering and management (operations research) from Oklahoma State University.

ALYSON G. WILSON is a research staff member at the Science and Technology Policy Institute of the Institute for Defense Analyses in Washington, DC. Previously, she was with the Department of Statistics at Iowa State University, a scientist at Los Alamos National Laboratory, and she was a biomedical researcher at the National Institutes of Health. She also served as a senior statistician and operations research analyst with Cowboy Programming Resources, where she planned, executed, and analyzed U.S. Army air defense artillery operational evaluations. Her research interests include reliability and information combination in scientific problems, Bayesian methods, and the application of statistics to problems in defense and national security. She has served on numerous national panels, including the Sandia National Laboratories' Predictive Engineering Science Panel. She is a fellow of the American Statistical Association. She holds a B.A. in mathematical sciences from Rice University, an M.S. in statistics from Carnegie Mellon University, and a Ph.D. in statistics from Duke University.

COMMITTEE ON NATIONAL STATISTICS